The Biology of Spermatozoa

Proceedings of the INSERM International Symposium, Nouzilly, November 4–7, 1973

The Biology of Spermatozoa

Transport, Survival and Fertilizing Ability

Editors: *E.S.E. Hafez*, Detroit, Mich. and *C.G. Thibault*, Jouy-en-Josas

78 figures, 50 tables, 1975

S. Karger · Basel · München · Paris · London · New York · Sydney

Cataloging in Publication
The biology of spermatozoa; transport, survival and fertilizing ability
Editors: E.S.E. Hafez and C.G. Thibault. Basel, New York, Karger (c1975)
Spermatozoa – physiology – congresses
I. Hafez, Elsayed Saad Eldin, 1922, ed. II. Thibault, C., ed. III. Institut national de la santé et de la recherche médicale IV. Series
Proceedings of the INSERM international symposium, Nouzilly, Nov. 4–7, 1973
ISBN 3–8055–2104–9

Printed in Switzerland by Thür AG Offsetdruck, Pratteln

Contents

Practical Aspects

Preface

Publication of this volume comes at a time when there is a need to clarify the problems related to sperm transport and survival; such problems are of great fundamental importance due to the use of synthetic steroids and IUD in the regulation of reproduction in the human female and in domestic mammals. General diversified forms of vertebrae are characterized by prolonged spermatozoa survival in the female genital tracts ranging from the salamander, hare, to the bat. Parameters of reproduction are presented here, for the first time, in a condensed form. Physiological and biochemical parameters of these species can contribute very richly to our knowledge of survival of mammalian spermatozoa.

E.S.E. Hafez, Detroit, Mich.
C.G. Thibault, Jouy-en-Josas

Sperm Transport in the Male

The Biology of Spermatozoa
INSERM Int. Symp., Nouzilly 1973, pp. 2–9 (Karger, Basel 1975)

Fluid Secretion by the Testis in the Transport and Survival of Spermatozoa

B.P. Setchell and G.M.H. Waites

ARC Institute of Animal Physiology, Babraham, Cambridge, and Department of Physiology and Biochemistry, The University, Whiteknights, Reading, Berks.

Introduction

Fluid leaves the testis through the efferent ducts at a rate of between 5 and 40 μl/g testis/h, faster in the smaller species, and normally contains between 0.3×10^8 and 1×10^8 spermatozoa per ml (21, 32). The spermatozoa are virtually immotile and incapable of fertilizing ova (18, 27). The flow of fluid from the testis is continuous and relatively steady and is much greater than the amount of material passing into the vas deferens, even during maximal sexual activity (39), because much of the fluid leaving the testis is reabsorbed in the epididymis.

Rete testis fluid is different from blood plasma or testicular lymph (23). Compared with blood plasma there is about 3 times as much potassium in RTF, and differences in all the other major ions; there is practically no glucose or fructose in RTF, but high concentrations of inositol; very little protein but many different proteins (13–16), less of most of the free amino acids, but substantially more of some (26, 28, 32), and appreciable amounts of testosterone, exceeding the concentration in arterial blood but less than in blood from the internal spermatic vein (5).

However, the fluid (STF) in the seminiferous tubules is different from rete testis fluid especially in its ionic (19, 22, 32) and protein composition (14–17) and there is evidence that there are two distinct secretions in the testis; a primary secretion, rich in potassium and bicarbonate, formed in the seminiferous tubules, and a secondary secretion, plasma-like in its ionic composition, secreted in relatively greater amounts either in the rete testis or in the tubuli recti. Furthermore, the concentration of spermatozoa, expressed as a percentage of the fluid, was greater in samples of STF than in RTF (19, 32) and the concentration of spermatozoa in the fluid retained after efferent duct ligation (EDL), was appreciably greater than the concentration in RTF (25).

Testicular Fluid and the Transport of Spermatozoa

Spermiation

Although hormone-induced sperm release is achieved in the toad by the development of fluid-filled vacuoles in the supporting cells near the spermatozoa (2), a similar mechanism, suggested to operate in the hamster with LH as the sperm-releasing hormone (33, 34), was not confirmed when fluid flow in either hamsters or rats was measured (25). Furthermore, *Fawcett* (10) has suggested instead that the sperm of mammals are released due to active movements of the apical region of the Sertoli cells, while the residual bodies are retained in the epithelium.

Fluid secretion begins in young animals before the first spermatozoa are shed (22, 24) and continues at normal or near normal rates in adults whose testes have been heated to reduce drastically the number of spermatozoa leaving the testis. It seems that sperm release and fluid secretion in mammals are not necessarily associated.

Transport of Spermatozoa out of the Seminiferous Tubules

The seminiferous tubules are long, double-ended tubes opening at both ends into the rete testis via the short tubuli recti (4, 20). Presumably, spermatozoa leave the tubules in a current of fluid, and the simplest situation would be that fluid is secreted in the tubule and emerges continually from both ends. However, this is not compatible with the observed differences in composition between newly secreted or 'primary' fluid and the fluid normally present in the tubule (32); these differences suggest instead that fluid is drawn into the tubules from the rete testis. Droplets of oil left in the seminiferous tubules are moved along the tubule out of the field of vision, as if carried by a current of fluid (32) and markers injected into the rete testis often pass into some tubules (3).

Isolated seminiferous tubules are capable of slow peristaltic activity and are sensitive to oxytocin. Smooth muscle cells also are present in the capsule of the testis which contracts spontaneously in the rat, rabbit and human – see *Setchell and Waites* (24) for references. Such contractions cause pressure changes within the testis and may provide a pumping action assisting the movement of fluid and spermatozoa to the epididymis.

When a focal lesion was produced in the testis by a fine hot needle, degeneration subsequently spread along the affected tubule but only in one direction (31). It spread along the shorter portion of the tubule to the rete, with little or no degeneration along the tubule in the other direction. This observation emphasizes the possible nutritive or regulatory roles of the fluid, and since no spermatozoa accumulate in the tubule, fluid produced in the tubules is sufficient to carry away the spermatozoa.

Transport of Spermatozoa from the Testis to the Epididymis

The flow of sperm-laden fluid from the rete is a continuous process, with no appreciable circadian rhythm of fluid flow. The lining of the rete testis has no contractile component and therefore the flow through the efferent ducts must be due to *vis a tergo* pressure, perhaps assisted by contractions of the testicular capsule (see above) or the single, short cilia which some of the cells have (8).

At some point, the lining of the rete and the efferent ducts changes quite rapidly from a functionless or possibly a secretory epithelium to a resorptive epithelium, possibly at the point where the two tissues of different embryological origin meet (20). At about the point where the resorptive function is maximal, the epididymal duct develops a muscular coat (1) which presumably helps to move the epididymal contents along the duct.

Effects of Testicular Fluids on the Survival of Spermatozoa

Once the spermatozoa leave the Sertoli cells, they are dependent on their endogenous reserves and on substances in the tubular and rete testis fluids.

Possible Exogenous Substrates in RTF

The testis depends heavily on glucose for its oxidative metabolism but this cannot apply to the spermatozoa because glucose is absent from RTF (21). Inositol in the RTF seems to be oxidized by the sperm to only a small extent. However, the spermatozoa contain endogenous inositol which may be utilized in preference to that in the fluid (36, 37).

A search for exogenous substrates has been made by measuring oxygen uptake of testicular spermatozoa, incubated either in RTF, or in an ionic mixture like RTF and buffered with bicarbonate-CO_2. Under these circumstances, the oxygen uptake of both ram and boar spermatozoa in natural RTF was appreciably higher than in the ionic mixture. Earlier comparisons are not valid because the respiration of the spermatozoa in RTF was contrasted with that of sperm in a phosphate-buffered saline solution resembling plasma, and not RTF (8a).

Lactate and acetate were the only substances of those added which restored the oxygen uptake of ram sperm in buffer to that in RTF, but fructose and pyruvate in abnormally high concentrations were also effective. With pig sperm, lactate was without effect but acetate did stimulate oxygen uptake. Inositol tended to decrease oxygen uptake slightly in both species. Furthermore, RTF spared the decrease in phospholipids which normally occurs when sperm are incubated in saline mixtures; the decrease in the most abundant fatty acid (22:6) is much more marked than the decrease in the next most abundant fatty acid (16:0). Lactate was without effect on the decrease in phospholipids (8a).

Testicular spermatozoa have been shown to be active in removing lipids from suspensions and metabolizing some of them (8a). There is only a small amount of lipid in RTF or in the caput epididymidis, but it may become important in the corpus and cauda epididymidis, where secretion of lipid, e.g. glyceryl phosphorylcholine, occurs (7).

Utilization of Endogenous Substrates by Testicular Spermatozoa

Ram testicular spermatozoa incubated in phosphate buffered saline mixtures metabolized a small fraction of their phospholipid content. However, spermatozoa collected at room temperature or at 15 or 4 °C have already used a large part of their triglyceride lipid reserves as was revealed when spermatozoa were collected directly into chloroform-methanol.

The temperature of collection also affects the composition of the remaining lipid. The phospholipid content of testicular spermatozoa collected at 0 and 37 °C is similar to the phospholipid content of spermatozoa collected directly into chloroform-methanol. However, at intervening collection temperatures, the phospholipid content is lower, and the content of most of the fatty acids obtained from the phospholipids decreases to reach a minimum at a collection temperature of 20 °C. This suggests that the fatty acids of the phospholipid molecule are more readily metabolized than the glycerol skeleton (8a).

Inhibitor of Acrosin

The type of inhibitor of acrosin or acrosomal proteinase known to be present in ejaculated semen is also present in RTF (29). The inhibitor in ram RTF interferes with *in vitro* fertilization (30) and three separate inhibitors have been found. None of the inhibitors inhibited the activities of chymotrypsin, plasmin, Kallikrein, thrombin, urokinase or acrosomal hyaluronidase or neuroaminidase, and they all were four times more effective against acrosin than against pancreatic trypsin. All three inhibitors from RTF were polypeptides, molecular weight about 6,500, and their amino acid composition differed slightly from each other and appreciably from other inhibitors that have been described in other species (24). The inhibitor in RTF may be important in neutralizing the proteolytic activity of any spermatozoa degenerating in the epididymis, or in preventing spermatozoa from penetrating the cells lining the epididymal duct.

Steroids and Steroid-Binding Proteins in Rete Testis Fluid

Testosterone was found in rete testis fluid collected from rams and bulls. Rat rete testis fluid contains approximately twice the testosterone concentration found in ram fluid but the levels were always less than in spermatic and testicular venous blood collected simultaneously, the ratio being approximately 0.9 in rat and 0.2 in ram (5). This difference may result from species differences in the topographical relations of the Leydig cells with the interstitial lymphatics and the tubular wall (9).

In addition to testosterone, dehydroepiandrosterone was identified in ram RTF; in fluid from bulls, androstenedione was found but not dihydrotestosterone. Oestrogens are present in high concentrations in boar RTF. However, cholesterol is present in rat and ram RTF in only low concentrations – see *Setchell and Waites* (24) for references.

The origin of the steroids in rete testis fluid is still unknown, although the weight of evidence favours Leydig cells rather than a tubular source. There is evidence from rats that androgens produced in the interstitium would have ready and rapid access to the tubules (23). Exogenous testosterone and related steroids have been added to ram (38) and bull (35) testicular spermatozoa to examine their action on oxidative and glycolytic activities. The experiments should be repeated using physiological concentrations of testosterone (10–30 ng/ml in the presence of androgen-binding proteins such as those which have been detected in RTF (11) and inferred from studies on testis and caput epididymis contents before and after efferent duct ligation (6, 12).

Conclusions

Fluid has been removed for analysis from the seminiferous tubules and also from the rete testis. These two fluids are different in ion and protein composition and also in the concentration of spermatozoa, and it seems probable that two different fluids are secreted, one by the seminiferous tubules and one by the tubuli recti or the epithelium lining the rete testis. Fluid secretion is probably not involved in the release of spermatozoa from the germinal epithelium, but is almost certainly involved in the transport of spermatozoa out of the seminiferous tubules and from the rete testis into the epididymis. The fluid also probably provides substrate for the spermatozoa to metabolize, although they may also draw on their endogenous reserves, particularly of lipid, and the fluid may be important in controlling this endogenous metabolism of the spermatozoa.

Rete testis fluid and epididymal seminal plasma contain appreciable concentrations of inhibitors of acrosomal proteinase, and although these inhibitors are not tightly bound to the spermatozoa, they may be important in controlling proteolytic activity during the maturation of the spermatozoa in the epididymis.

Finally, although the concentrations of testosterone and related steroids in RTF are lower than in testicular venous blood, nevertheless they are much higher than in fluids bathing cells elsewhere in the body, and the presence of high affinity androgen binding proteins may ensure their transport to receptor sites in the epididymal lumen. The steroids in the fluid bathing testicular spermatozoa may influence their metabolism.

References

1 *Baumgarten, H.G.; Holstein, A.F., and Rosengren, E.:* Arrangement, ultrastructure and adrenergic innervation of smooth musculature of the ductuli efferentes, ductus epididymidis and ductus deferens of man. Z. Zellforsch. *120:* 37–39 (1971).

2 *Burgos, M.H. and Vitale-Calpe, R.:* The mechanism of spermiation in the toad. Amer. J. Anat. *120:* 227–251 (1967).

3 *Clegg, E.J. and MacMillan, E.W.:* The uptake of vital dyes and particulate matter by the Sertoli cells of the rat testis. J. Anat., Lond. *99:* 219–229 (1965).

4 *Clermont, Y. and Huckins, C.:* Microscopic anatomy of the sex cords and seminiferous tubules in growing and adult male albino rats. Amer. J. Anat. *108:* 79–97 (1961).

5 *Cooper, T.J. and Waites, G.M.H.:* Testosterone in rete testis fluid and blood of rams and rats. J. Endocrin. (in press, 1975).

6 *Danzo, B.J.; Orgebin-Crist, M.-C., and Toft, D.O.:* Characterization of a cytoplasmic receptor for 5α-dihydrotestosterone in the caput epididymis of intact rabbits. Endocrinology *92:* 310–317 (1973).

7 *Dawson, R.M.C. and Rowlands, I.W.:* Glycerylphosphorylcholine in the male reproductive organs of rats and guinea pigs. Quart. J. exp. Physiol. *44:* 26–34 (1959).

8 *Dym, M.:* The mammalian rete testis. Anat. Rec. *172:* 304 (1972).

8a *Evans, R.W. and Setchell, B.P.:* The effect of rete testis fluid on the metabolism of testicular spermatozoa. J. Reprod. Fertil. (in press, 1975).

9 *Fawcett, D.W.:* Observations on the organization of the interstitial tissue of the testis and on the occluding cell junctions in the seminiferous epithelium. Adv. Biosci. *10:* 83–99 (1973).

10 *Fawcett, D.W.:* The ultrastructure and functions of the Sertoli cell; in American handbook of physiology, section on endocrinology – Male reproduction (in press).

11 *French, F.S. and Ritzen, E.M.:* Androgen-binding protein in efferent duct fluid of rat testis. J. Reprod. Fertil. *32:* 479–483 (1973).

12 *French, F.S. and Ritzen, E.M.:* The high affinity androgen binding protein in rat testis: evidence for secretion into efferent duct fluid and absorption by epididymis. Endocrinology *93:* 88–95 (1973).

13 *Johnson, M.H. and Setchell, B.P.:* Protein and immunoglobin content of rete testis fluid of rams. J. Reprod. Fertil. *17:* 403–406 (1968).

14 *Kormano, M.; Koskimies, A.I., and Hunter, R.L.:* The presence of specific proteins in the absence of many serum proteins in the rat seminiferous tubule fluid. Experientia *27:* 1461–1463 (1971).

15 *Koskimies, A.I. and Kormano, M.:* The proteins in fluids from the seminiferous tubules and rete testis of the rat. J. Reprod. Fertil. *34:* 433–444 (1973).

16 *Koskimies, A.I.; Kormano, M., and Alfthan, O.:* Proteins of the seminiferous tubule fluid in man – evidence for a blood-testis barrier. J. Reprod. Fertil. *32:* 79–86 (1973).

17 *Koskimies, A.I.; Kormano, M., and Lahti, A.:* A difference in the immunoglobulin content of seminiferous tubule fluid and rete testis fluid of the rat. J. Reprod. Fertil. *27:* 463–465 (1971).

18 *Lambiase, J.T. and Amann, R.P.:* Infertility of rabbit testicular spermatozoa collected in their native fluid environment. Fertil. Steril. *24:* 65–67 (1973).

19 *Levine, N. and Marsh, D.J.:* Micropuncture study of electrochemical aspects of fluid and electrolyte transport in individual seminiferous tubules, the epididymis and vas deferens in rat. J. Physiol., Lond. *231:* 557–570 (1971).

20 *Roosen-Runge, E.C.:* The rete testis in the albino rat: its structure, development and morphological significance. Acta anat. *45:* 1–30 (1961).

21 *Setchell, B.P.:* Testicular blood supply, lymphatic drainage and secretion of fluid; in *Johnson, Gomes and van Demark* The testis, chap. 3, vol. 1 (Academic Press, New York 1970).

22 *Setchell, B.P.:* The secretion of fluid by the testes of rats, rams and goats with some observations on the effect of age, cryptorchidism and hypophysectomy. J. Reprod. Fertil. *23:* 79–85 (1970).

23 *Setchell, B.P. and Waites, G.M.H.:* The blood-testis barrier; in American handbook of physiology, section on endocrinology. Volume on reproductive biology of the male (in press).

24 *Setchell, B.P. and Waites, G.M.H.:* Fluid secretion by the testis in the transport and survival of spermatozoa; in *Thibault and Hafez* Transport et survie et pouvoir fécondant des spermatozoïdes chez les vertébrés (Inserm, Paris 1974).

25 *Setchell, B.P.; Duggan, M.C., and Evans, R.W.:* The effects of gonadotrophins on fluid secretion and sperm production by the rat and hamster testis. J. Endocrin. *56:* 27–36 (1972).

26 *Setchell, B.P.; Hinks, N.T.; Voglmayr, J.K., and Scott, T.W.:* Amino acids in ram testicular fluid and semen and their metabolism by spermatozoa. Biochem. J. *105:* 1061–1065 (1967).

27 *Setchell, B.P.; Scott, T.W.; Voglmayr, J.K., and Waites, G.M.H.:* Characteristics of testicular spermatozoa and the fluid which transports them into the epididymis. Biol. Reprod. Suppl. *1:* 40–66 (1969).

28 *Sexton, T.J.; Amann, R.P., and Flipse, R.J.:* Free amino acids and protein in rete testis fluid, vas deferens plasma, accessory sex gland fluid and seminal plasma of the conscious bull. J. Dairy Sci. *52:* 412–416 (1971).

29 *Suominen, J. and Setchell, B.P.:* Enzymes and trypsin inhibitor in the rete testis fluid of rams and boars. J. Reprod. Fertil. *30:* 235–245 (1972).

30 *Suominen, J.; Kaufman, M.H., and Setchell, B.P.:* Prevention of fertilization *in vitro* by the acrosin inhibitor from rete testis fluid of the ram. J. Reprod. Fertil. *34:* 385–388 (1973).
31 *Suoranto, H.:* Tubular damage caused by local thermal injury or microembolism of the rat testis. Virchows Arch. Abt. B. Zellpath. *8:* 299–308 (1971).
32 *Tuck, R.R.; Setchell, B.P.; Waites, G.M.H., and Young, J.A.:* The composition of fluid collected by micropuncture and catheterization from the seminiferous tubules and rete testis of rats. Pflügers Arch. ges. Physiol. *318:* 225–243 (1970).
33 *Vitale-Calpe, R. and Burgos, M.H.:* The mechanism of spermiation in the hamster. I. Ultrastructure of spontaneous spermiation. J. Ultrastruct. Res. *31:* 381–393 (1970).
34 *Vitale-Calpe, R. and Burgos, M.H.:* The mechanism of spermiation in the hamster. II. The ultrastructural effects of coitus and of LH administration. J. Ultrastruct. Res. *31:* 394–406 (1970).
35 *Voglmayr, J.K.:* The effect of testosterone and related steroids on the oxidative and glycolytic activity and on lipid synthesis of testicular and ejaculated bull spermatozoa. Acta endocrin., Kbh. *68:* 793–804 (1971).
36 *Voglmayr, J.K. and Amann, R.P.:* The distribution of free myoinositol in fluids, spermatozoa and tissues of the bull genital tract and observations on its uptake by the rabbit epididymis. Biol. Reprod. *8:* 504–513 (1973).
37 *Voglmayr, J.K. and White, I.G.:* Synthesis and metabolism of myoinositol in testicular and ejaculated spermatozoa of the ram. J. Reprod. Fertil. *24:* 29–37 (1971).
38 *Voglmayr, J.K.; Murdoch, R.N., and White, I.G.:* Metabolism of ram testicular spermatozoa in the presence of testosterone and related steroids. Acta endocrin., Kbh. *65:* 565–576 (1970).
39 *Waites, G.M.H. and Setchell, B.P.:* Some physiological aspects of the function of the testis; in *McKerns* The gonads, chap. 21 (Appleton Century Crofts, New York 1969).

Authors' addresses: Dr. *B.P. Setchell,* Agricultural Research Council, Institute of Animal Physiology, *Babraham CB2 4AT;* Prof. *G.M.H. Waites,* Department of Physiology and Biochemistry, The University, Whiteknights, *Reading RG6 2AJ* (England)

The Biology of Spermatozoa
INSERM Int. Symp., Nouzilly 1973, pp. 10–16 (Karger, Basel 1975)

Sperm Transport and Survival in Male Birds

Michel de Reviers

Institut National de la Recherche Agronomique, Station de Recherches Avicoles, Nouzilly

Introduction

In birds, the origin and composition of seminal plasma (12), and its interaction with spermatozoa present original features because of the absence of a well-developed epididymis and individualized accessory glands (11). This paper concerns the maturation, survival and transport of sperm in male birds.

Materials and Methods

The cockerels used in this work were raised in individual cages and were exposed to natural daylight.

Timing of Sperm Transport in the Excurrent Ducts of the Testis

23 White Leghorn cockerels were injected individually with 2 mCi of ^{3}H-thymidine during March-April or August. Six of them were slaughtered on day 12 after injection and their testes were prepared for histology. The 17 remaining cockerels were ejaculated daily (4), either from the day of injection or from day 10. This delay induced no observable variations in the results. Testis slides and sperm smears were autoradiographed by dipping in Ilford K5 photographic emulsion.

Extragonadal Sperm Reserve Evaluation

Two experiments were conducted to determine the number and the distribution of spermatozoa in the excurrent ducts of the testis, the effect of sperm collection and the effect of season. The first two points were studied during October and November, using 12 nonejaculated cocks and 8 ejaculated daily for 25 days.

Seasonal variations of the extragonadal sperm reserve were determined by comparison with the results obtained either during March-April (14 males) or during August (16 males) from cockerels ejaculated daily for 17 or 27 days.

In these two experiments, the excurrent ducts were split into 4 parts: the epididymal region, the proximal and distal halves of the deferent duct, and the spermatic vesicle. After homogenization in sucrose, the number of spermatozoa was determined with hemocytometers (6–10 grids per part). A small piece of each testis was treated in the same way. Both elongated spermatids and spermatozoa were then counted; their total number is referred to as 'testicular number' (18).

Frequency of Semen Collection and Sperm Motility

This work was done with the collaboration of *Petitjean* (21) with 14- to 27-week-old cockerels. Sperm collections were taken once a week until the cockerels were 23 weeks old, and then once a day. Each ejaculate was diluted (v/v) in a citrate diluent and stored for 5 h at 4 °C. The percentage of motile sperm was then measured with an electronic device built in Petitjean's laboratory.

Results

Timing of Sperm Transport in the Excurrent Ducts of the Testis

Previous results (20) have shown that meiotic prophase lasts 5.5 days and spermiogenesis 7 days, including 4.5 days for elongated spermatids. Therefore, the daily sperm production (DSP) is 'testicular number'/4.5.

In the present work on the testis, the first labeled spermatozoa were released out of the seminiferous epithelium from days 12 to 13 after injection of the marker. This timing is the same in every cockerel independent of the season. In the daily ejaculates, however, the appearance of the first labeled spermatozoa occurred on days 13–14 during August, or on day 25 during March and April (fig. 1).

So, while the total duration of the meiotic prophase and spermiogenesis seem to be constant in the cockerel, the time of sperm transport shows seasonal variations, being 24 h or less during August and 48 h during March and April.

The evolution of the labeling index[1] shows that we observe two consecutive waves of labeled spermatozoa (fig. 2). In August, these two waves peak sharply at a distance of 2–3 days in 8 out of 11 cockerels examined. However, in March-April, these two waves are spread over 2–4 days at a distance of 4–5 days. These results (spread of the waves and their delay from the injection time) are interpreted as the manifestation of a slowing of the sperm transport in springtime. The particular case of a single peak may be considered as an artifact due to the frequency of sperm collection, too low to allow two peaks to appear.

Quantification of Extragonadal Sperm Reserve (ESR)

In nonejaculated cockerels, the mean ESR were 6.8×10^9 spermatozoa (extremes: 0.9 to 19.0 billions per cockerel). About 65 % of the ESR were

1 Labeling index = number of labeled spermatozoa × 100/total number of observed spermatozoa.

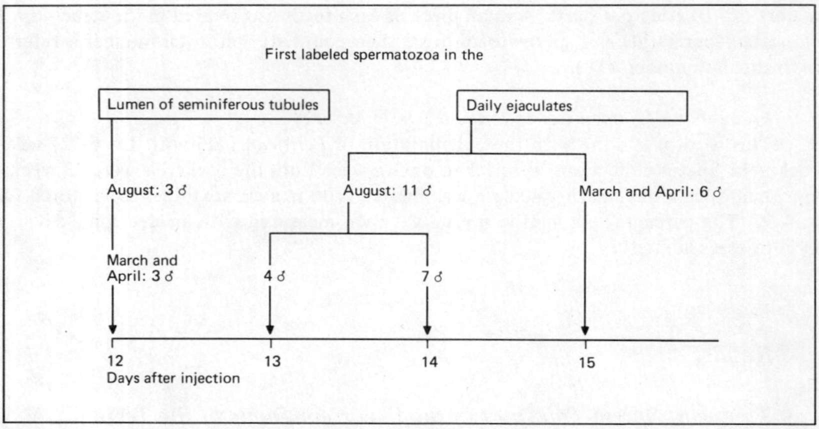

Fig. 1. Timing of the first labeled spermatozoa in the testes and in the daily ejaculates obtained from cockerels after 2 mCi ^{3}H-thymidine injections done on the day O either during March-April or August. Note the seasonal variations of the sperm transit duration.

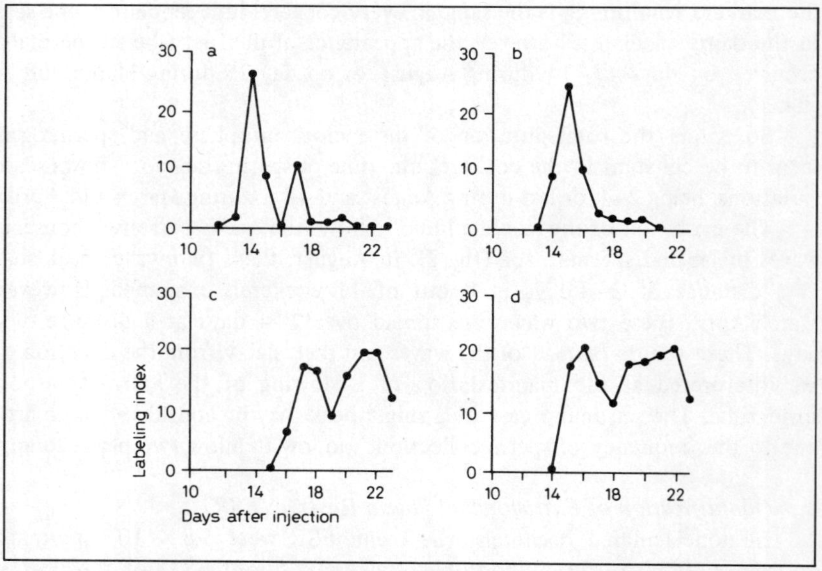

Fig. 2. The four patterns of the labeling index of the spermatozoa in the daily ejaculates obtained from cockerels after 2 mCi ^{3}H-thymidine injection done on the day O either during August (a, b) or March-April (c, d). Note the presence of two successive waves of labeled spermatozoa (a, c, d). They might be superimposed in b (see text). Every point of each curve has been obtained after observation of 1,000–2,000 spermatozoa.

located in the distal half of the deferent duct. The epididymis contained only 5 % of the total ESR. The daily testicular sperm production (DSP) was about 2×10^9 spermatozoa in the nonejaculated cockerels. Therefore, the ESR represents the equivalent of 3.5 days of the testicular sperm production.

In the cockerels ejaculated daily, the total ESR were higher during March and April (4.5×10^9 spermatozoa) than during August (1.6×10^9). The most important relative variation of the reserves occurred in the proximal half of the deferent ducts. These ESR variations correspond to parallel variations of testicular weight (13 versus 8.8 g, respectively) and of DSP (2.1 versus 1.5×10^9 spermatozoa). Therefore, the total ESR in cockerels ejaculated daily represent about 2 days of the DSP during March and April versus 1 day during August.

Sperm Collection and Motility of Ejaculated Spermatozoa

In young cockerels, the percentage of motile sperm collected once a week, measured after dilution and 5 h storage at 4 °C, increases from the 14th to the 17th week of age. This percentage is relatively constant (15–20 %) until week 22. When the frequency of semen collection changes from once a week to once a day, the percentage of motile sperm increases up to 25–30 % to reach a new plateau. This difference is significant at the 5 % level. It can be concluded that by increasing the frequency of sperm collection, the motility of ejaculated sperm after storage increases.

Discussion

When taken directly in the testis, the spermatozoa of the cockerel and the turkey, as well as those of mammals, exhibit low motility and have a poor fertilizing rate. In birds, these two parameters become normal in the distal half of the deferent duct (14, 23). During this maturation process, no evident morphological modifications occur except a thickening of the inner mitochondrial membrane (6). The maturation of bird sperm might be associated with the lipomucoprotein and glycoprotein secretions which have been observed in the male genital tract (11, 12). We assume that the interaction between these secretions and spermatozoa could result in the acquisition of a protective coating by the spermatozoa, according to *Gunn* (7), or to an internal interaction with nucleoproteins. Although such mechanisms are still not proven in birds, they seem more likely than the aging process postulated by *Young* (29).

In the male genital tract of the cockerel, spermatozoa and India ink particles seem to migrate at the same speed in the deferent ducts (14, 20, 24). Sperm transport would therefore be a passive mechanism involving strereocilia or, more likely, a fluid flow from the testis and peristalsis of the duct walls.

Is every spermatozoa of a given generation transported at the same speed in the deferent ducts of birds? The answer to this question is difficult. In the cockerel, when two peaks were evident in the labeling index, we determined the number of silver grains contained in the photographic emulsion above the labeled spermatozoa. If these two peaks corresponded to consecutive sperm generations, the ratio between the mean (or modal) numbers of silver grains in the first and second peaks would be approximately 2 to 1. Unexpectedly, it is quite variable from one cockerel to another (0.65 to 1.60). Such a situation has been observed in the quail (1) and does not appear to be an artifact. This leads to other questions related to the duration of DNA synthesis in bird spermatocytes and spermatogonia, as well as to the synchronization of the spermatogenic processes and sperm transport in the male genital tract of birds.

Our results show that the transit time of spermatozoa in the deferent ducts of the cockerel varies according to the season, being maximum (respectively minimum) when the DSP is highest (respectively lowest). We do not know the reason why. Perhaps the capacity of the extragonadal reservoir is modified by the season. This point, difficult to verify in the cockerel, is quite evident in passerine birds. In these birds, the deferent ducts are well developed only during the breeding season, forming a glomus, the volume reaching one third of the testis volume (2, 22, 28). This enormous enlargement of the deferent ducts is testosterone-controlled.

By contrast, testosterone does not prolong the life of spermatozoa in the deferent ducts of the castrated cock (15), while it does in mammals (3, 10, 13, 17). Another interesting fact is that the drug α-chlorohydrin, which is an epididymal chemosterilant in the rat (6) and an inhibitor of selective resorption of the testicular fluid, has no effect on fertility in the male quail (9). We can thus assume that the physiology of the male genital tract in birds is rather different than that in mammals.

The results obtained in timing sperm transport in the cockerel agree with our data on extragonadal sperm reserves. When θ is the total duration of sperm transport, the DSP and the ESR verify the following relation: $ESR = DSP \times \theta$.

Even in nonejaculated cockerels, the mean ESR does not exceed 3.5 days of the DSP. The latter, numbering 2 billions, makes it obvious that many spermatozoa are eliminated after 3 days of 'sexual rest'. How does this elimination occur? Probably not by resorption. This process can be effectively observed in the cockerel; it is quantitatively important only when the deferent ducts are ligated for a long period (25). Finally, we think that the main process of sperm elimination in cockerels is either due to spontaneous ejaculation or to a continuous expulsion with urine and feces. It must be emphasized that these processes are less important when sperm is collected once a day instead of once every two days. From 0 to 25 % of the DSP is lost in the first case, and 50 % in the second case (21).

From the comparison between results obtained with India ink particles and those with labeled spermatozoa, we conclude that sperm collection (done according to the massage method) does increase the rate of sperm transit through the excurrent ducts. As a matter of fact, this transit lasts 24–48 h in cockerels ejaculated daily (14, 20) versus 3–4 days in nonejaculated cockerels (20, 24). While the quality and the fertilizing ability of the spermatozoa increase with the frequency of semen collection, reciprocally (12, 19, 21) it may be concluded for the cockerel that: (1) after the minimum time necessary for maturation, spermatozoa progressively lose their fertilizing ability when they stay in the male genital tract, and (2) when they are stored in the male genital tract, spermatozoa lose their fertilizing ability a long time before they die.

Sperm survival does not necessarily imply conservation of fertilizing ability; these two characteristics must be distinguished *in vivo* and *in vitro*. These conclusions can be also drawn from the data reported in mammals (5, 26, 27).

Finally, as far as glandular cells, sperm maturation, sperm survival and extragonadal sperm reserves are concerned, we conclude that the avian analogy of the mammalian epididymis is the deferent duct.

References

1 *Amir, D.; Braun-Eilon, B., and Schindler, H.:* Passage and disappearance of labelled spermatozoa in the genital tract of the male japanese quail in segregation or cohabitation. Ann. Biol. anim. *13:* 321–328 (1973).

2 *Bailey, R.E.:* Accessory reproductive organs of male Fringillid birds: seasonal variations and response to various sex hormones. Anat. Rec. *115:* 1–19 (1953).

3 *Benoit, J.:* Recherches anatomiques, cytologiques et histophysiologiques sur les voies excrétrices du testicule chez les mammifères; D.Sc. thesis, Strasbourg (1925).

4 *Burrows, W.H. and Quinn, J.P.:* The collection of spermatozoa from the domestic fowl and turkey. Poult. Sci. *16:* 19–24 (1937).

5 *Groot, B. de:* Ageing of bull sperm at the end of the passage through the epididymis. J. Reprod. Fertil. *2:* 107–116 (1961).

6 *Ericsson, R.J. and Baker, W.F.:* Male antifertility compounds: biological properties of U-5897 and U-15,646. J. Reprod. Fertil. *21:* 267 (1970).

7 *Gunn, R.M.C.:* Fertility in sheep. Artificial production of seminal ejaculation and the characters of ejaculated spermatozoa contained therein. Austr. Comm. Res. Bull. *94:* 1–116 (1936).

8 *Gunn, S.A. and Gould, T.C.:* Specificity of the vascular system of the male reproductive tract. J. Reprod. Fertil. Suppl. *10:* 75–95 (1970).

9 *Jones, P.; Kominkova, E., and Jackson, H.:* Effects of antifertility substances on male japanese quail. J. Reprod. Fertil. *29:* 71–78 (1972).

10 *Kirillov, V.S. and Morozov, V.A.:* Sperm survival duration in epididymes isolated from testes (in russian). Usp. Zootekn. Nauk. *2:* 19–22 (1927).

11 *Lake, P.E.:* The male reproductive tract of the fowl. J. Anat., Lond. *91:* 116–129 (1957).

12 *Lake, P.E.:* The male in reproduction; in *Bell and Freeman* Physiology and biochemistry of the domestic fowl, vol. 3 (Academic Press, New York 1971).
13 *Moore, C.R.:* Sperm activity and testis hormone. J. exp. Zool. *50:* 455 (1928).
14 *Munro, S.S.:* Functional changes in fowl sperm during their passage through the excurrent ducts of the fowl. J. exp. Zool. *79:* 71–92 (1938).
15 *Munro, S.S.:* The effects of testis hormones on the preservation of sperm life in the vas deferens of the fowl. J. exp. Biol. *15:* 186–196 (1938).
16 *Nicander, L. and Hellström, B.:* Increased thickness of the inner mitochondrial membrane during sperm maturation in the domestic rooster. Exp. Cell Res. *48:* 622 (1967).
17 *Orgebin-Crist, M.C.:* Transport, survival and maturation of epididymal sperm in rodents. Int. Symp. on Sperm Transport, Survival and Fertilizing Ability, Nouzilly 1973 (in press).
18 *Ortavant, R.:* Le cycle spermatogénétique chez le bélier; D.Sc. thesis, Paris (1958).
19 *Petitjean, M.J.:* Résultats expérimentaux sur la subfertilité liée à la crête rosacée chez le coq. 14th World's Poultry Congr., Madrid 1970.
20 *Reviers, M. de:* Détermination de la durée des processus spermatogénétiques chez le coq à l'aide de la thymidine tritiée. 6th Congr. Reprod. anim. Insem. artif., vol. 1, pp. 183–185 (Paris 1968).
21 *Reviers, M. de et Petitjean, M.J.:* Effets du gène de nanisme dw sur la production de spermatozoïdes chez le coq en croissance. Ann. Génét. Sél. anim. *3:* 313–321 (1973).
22 *Riddle, O.:* The cyclical growth of the vesicula seminalis in birds is hormone controlled. Anat. Rec. *37:* 1–12 (1927).
23 *Saeki, Y. and Brown, K.I.:* Fertilizing ability and laboratory evaluation of ejaculated and vasa deferentia semen in turkeys. Poult. Sci. *51:* 905–909 (1962).
24 *Takeda, A.:* Labelling of cock's spermatozoa with radioactive phosphate. Jap. J. zootech. Sci. *40:* 412 (1969).
25 *Tingari, M.D. and Lake, P.E.:* Ultrastructural evidence for resorption of spermatozoa in the excurrent ducts of the testis of the fowl *(Gallus domesticus).* J. Reprod. Fertil. *31:* 373–381 (1972).
26 *Wales, R.G. and White, I.G.:* The susceptibility of spermatozoa to temperature shock. J. Endocrin. *19:* 211–220 (1959).
27 *Willet, E.L. and Ohms, J.I.:* Influence of seminal plasma on maturity of bovine spermatozoa upon their freezability. J. Dairy Sci. *41:* 90–95 (1958).
28 *Wolfson, A.:* Sperm storage at lower-than-body temperature outside the body cavity in some passerine birds. Science *120:* 68–71 (1954).
29 *Young, W.C.:* A study of the function of the epididymis. I. Is the attainment of full spermatozoon activity attributable to some specific action of the epididymal secretion? J. Morph. *47:* 479–495 (1929).

Author's address: *M. de Reviers,* Institut National de la Recherche Agronomique, Station de Recherches Avicoles, *Nouzilly, F–37380 Monnaie* (France)

The Biology of Spermatozoa
INSERM Int. Symp., Nouzilly 1973, pp. 17–27 (Karger, Basel 1975)

Androgen Regulation of Epididymal Function

Benjamin J. Danzo, Marie-Claire Orgebin-Crist and Barbara C. Eller
Center for Population Research and Studies in Reproductive Biology, Vanderbilt University School of Medicine, Nashville, Tenn.

Introduction

The main functions of the epididymis are to serve as a conduit for the transport of spermatozoa from the testis to the ejaculatory duct and to provide a favorable environment for sperm maturation and storage prior to ejaculation. The morphological and functional integrity of the epididymis is androgen-dependent (3). Changes in tubular diameter, epithelial height, and histochemical characteristics after castration and their repair or prevention by the administration of exogenous testosterone have been extensively documented (1, 3, 9). The regressive changes in the epididymis following castration or hypophysectomy are associated with a loss of motility and death of the sperm contained therein (3, 29). It has been demonstrated recently that sperm maturation in the epididymis, i.e. the development of their ability to fertilize, and sperm survival are also androgen-dependent phenomena (11, 18, 20).

The generalized scheme for the mechanism of steroid hormone action appears to be (a) binding of the hormone to a cytoplasmic receptor (23), (b) entrance of the hormone-receptor complex into the nucleus where it interacts with chromatin (27), (c) the synthesis of specific proteins (19), and (d) subsequent events which were classically associated with the 'action' of a given hormone.

The first indication of androgen-binding moieties in the epididymis was obtained in studies examining uptake of labeled androgens by this tissue (5). Epididymal androgen-binding proteins (ABP) have recently been reported by several investigators (6, 16, 24). We have shown that such a protein exists in the caput epididymidis of sexually mature intact rabbits, but not in the corpus or cauda (10) and that a pattern of androgen-binding activity exists in the various epididymal segments (caput, corpus, and cauda) of sexually immature intact rabbits. Our recent studies have demonstrated a decrease or elimination of androgen binding in the epididymis following procedures that prevent testicular fluid from entering it. A similar finding has been made by *French and Ritzén* (14). Papers by *Hansson et al.* (17) and *Ritzén et al.* (25) have also provided data which show that an ABP is present in testis cytosol which is undistinguishable by several criteria from the ABP in cytosol prepared from the epididymis. These observations indicate that at least a portion of the ABP in the epididymis is derived from testicular fluid. In addition to the ABP of testicular origin, there is also evidence of another ABP which is derived from the epididymis *per se.* Epididymal ABPs can be demonstrated in the cytoplasm (6) and in the nuclei (7, 28). The studies reported here were undertaken to examine the effects of castration and other surgical procedures on androgen binding to cytosol prepared from mature rabbit epididymides; and, also, to determine if a correlation existed between testicular and epididymal maturation and androgen-binding activity in immature rabbit epididymides.

Results

For details of the methods and materials used in these studies, see *Danzo et al.* (10, 11).

The Effect of Castration on Binding of (^{3}H) 5αDHT to Epididymal Segments

A summary of the (^{3}H) 5αDHT studies on castrated animals is shown in table I. As can be noted, contrary to the results obtained with mature intact rabbits in which binding only occurred in the caput (10), many of the castrated animals exhibited (^{3}H) 5αDHT binding activity in all three segments. This phenomenon was particularly noticeable in animals that had been castrated for 3 or 4 days. Since the entire epididymis is dependent on the presence of androgen for its maintenance and function, one would anticipate the presence of androgen receptors along the length of the duct.

The Effect of Duct Ligation on (^{3}H) 5αDHT Binding to Cytosol from the Caput Epididymidis

Either 3 days or 2 weeks following unilateral efferent duct ligation, (^{3}H) 5αDHT binding to caput cytosol from the experimental and control side was examined. The testes of the duct-ligated and sham-operated rabbits were examined histologically to ascertain morphological changes that had occurred due to

Table I. Effect of castration of 5αDHT binding to epididymal cytosol

Number of rabbits	Duration of castration, days	Number of rabbits showing 5αDHT binding to		
		caput	corpus	cauda
25	0	25	0	0
7	1	6	7	4
6	2	6	4	3
8	3	8	8	8
10	4	10	10	10
6	5	5	2	4
6	6	4	1	2
10	7	7	4	7
8	14	6	4	5

1 Binding was determined by sucrose gradient ultracentrifugation.

the surgical procedures. The testis on the experimental side of duct-ligated animals showed the typical degenerative changes that result from accumulation of fluid within the seminiferous tubules; namely, the tubules had become distended and there was atrophy of the seminiferous epithelium. These changes were detectable 3 days after ligation, but were more marked after 2 weeks. We refer to testes of this type as showing the 'ligation effect'. In contrast, the testes on the sham-operated side were normal in appearance, indicating that the sham procedure did not interfere with the flow of testicular fluid and that the surgery *per se* did not cause any degenerative changes.

When binding of (^{3}H) 5αDHT to caput cytosol from five rabbits whose ductuli efferentes had been ligated for 3 days was examined by sucrose gradient ultracentrifugation, there was a decrease in (^{3}H) 5αDHT binding to caput cytosol from the ligated side as compared to the control side (table II). Two weeks following duct ligation, essentially no detectable (^{3}H) 5αDHT binding was present in caput cytosol from the ligated side (table III).

Two exceptions were noted in these studies: rabbit No. 6 had equal binding on the ligated and control side (table II), yet the testes showed the ligation effect indicating that most of the testicular fluid had been prevented from entering the epididymis. It is possible that after 3 days, all of the testicular fluid may not have yet been cleared from the caput. In the 2-week ligation series, there was also one rabbit whose testis on the ligated side was degenerated due to the accumulation of testicular fluid, but which exhibited near equal 5αDHT binding on the ligated and control sides (rabbit No. 21, table III). One would assume that two weeks after duct ligation all of the testicular fluid would have been cleared from the epididymis. Thus, it is unlikely that the binding was due

Table II. Effect of 3-day duct ligation on 5αDHT binding to caput cytosol

Rabbit No.	Binding (ligated side)	Binding (intact side)	Ligation effect on testis
1[1]	++++[2]	++++	no
2[1]	++++	++++	no
3[1]	++++	++++	no
4[1]	++++	++++	no
5	+	++++	yes
6	++++	++++	yes
7	+	++++	yes
8	+	++++	yes
9	+	++++	yes

1 Sham-operated controls.
2 The amount of binding is rated from '0' to '++++'.

Table III. Effect of 2-week duct ligation on 5αDHT binding to caput cytosol

Rabbit No.	Binding (ligated side)	Binding (intact side)	Ligation effect on testis
10[1]	++++[2]	++++	no
11[1]	++++	++++	no
12[1]	++++	++++	no
13[1]	++++	++++	no
14	+	++++	yes
15	0	++++	yes
16	0	++++	yes
17	0	++++	yes
18	0	++++	yes
19	0	++++	yes
20	+	++++	yes
21	++++	++++	yes

1 Sham-operated controls.
2 The amount of binding is rated from '0' to '++++'.

to testicular fluid remaining in the epididymis from the time of ligation. Since there was no decrease in binding following sham procedures, these results suggested that the decrease in binding to caput cytosol in duct-ligated rabbits was not due to surgical trauma incurred by caput tissue and suggested that the decrease was a consequence of preventing testicular fluid from entering the epididymis.

Effect of Hemicastration

To completely eliminate the source of possible testicular androgen binding proteins, we removed the testis from one side, leaving the epididymis intact. Three days later, binding of (^{3}H) 5αDHT on the hemicastrated side was greatly decreased or eliminated, whereas the amount of binding on the control side was normal. These results were obtained with five out of five animals. Hemicastration for two weeks completely eliminated (^{3}H) 5αDHT binding activity from caput cytosol on the experimental side. This observation was also made on five out of five rabbits. However, (^{3}H) 5αDHT binding was still present on the control side.

The Effect of Artificial Cryptorchidism

The decrease in testicular fluid secretion caused by elevated temperature has been demonstrated by several authors (2, 26). Unilateral artificial cryptorchidism was produced in a group of nine rabbits and 32–40 days later we prepared cytosol from the caput epididymidis on the cryptorchid and control sides. Caput cytosol from the epididymis from the scrotal testis showed a peak of macromolecular-bound (^{3}H) 5αDHT. However, no such peak was demonstrable in cytosol obtained from the caput which had been confined to the abdomen. This pattern was observed in eight out of nine animals. In one animal there was a 42-percent decrease of ABP on the cryptorchid side as compared to the scrotal side.

Histological examination of the cryptorchid testes revealed that the typical degenerative processes associated with cryptorchidism had occurred. The seminiferous rubules had atrophied and the only germinal elements present were the spermatogonia.

5αDHT Binding in Immature Rabbits

To determine whether the appearance of ABP in the epididymis correlated with testicular maturation and with fluid production by the testis, we untertook an analysis of (^{3}H) 5αDHT binding to epididymal segments from sexually immature intact rabbits of various ages.

Sucrose Gradient Analysis of Epididymal Cytosol

When cytosols from the caput, corpus, and cauda of rabbits (20–104 days of age) were analyzed for 5αDHT binding, a pattern of binding activity emerged. This pattern was: (a) animals in which there was little or no detectable binding

Table IV. Binding of (^{3}H) 5αDHT to immature rabbit epididymides

Number of rabbits	Group	Percentage of binding[1] caput	corpus	cauda	Age range, days	Body wt range, kg	Testis wt range, g
10[2]	I	28	trace[3]	56	20–30	0.5–0.7	0.01–0.04
10[2]	I	12	6	53	40–50	1.1–1.4	0.08–0.10
16[4]	I	trace	trace	trace	46–104	0.9–2.1	0.05–0.40
8[4]	II	132	80	122	60–104	1.8–2.5	0.40–0.90
10[4]	III	110	0	0	60–98	1.8–2.5	0.70–1.60

1 Percentage of the binding of (^{3}H) 5αDHT occurring to caput cytosol from mature rabbits.

2 These data were obtained using cytosol prepared from pooled epididymal segments from rabbits of the indicated age range.

3 Trace indicates that binding was virtually undetectable using sucrose gradient analysis.

4 Individual animals were used.

to any of the segments; (b) animals in which binding could clearly be demonstrated in all three segments, and (c) animals in which (^{3}H) 5αDHT binding could be demonstrated only in the caput epididymidis.

When cytosols from individual rabbits in the first group were examined for 5αDHT binding, there was little or no detectable binding (table IV). However, when cytosol was prepared from pools of epididymides (thus allowing us to increase the amount of cytosol included in the incubation mixture) from animals 20–30 and 40–50 days of age and analyzed for binding in a manner identical to the above, peaks of macromolecular bound hormone were detected. The epididymides from the 20- to 30-day-old rabbits showed a small but clear binding peak in the caput and cauda; the corpus showed only a trace of binding as was indicated by a shoulder on the gradient profile (table IV). Cytosols from all three segments of the 40- to 50-day-old rabbits showed small peaks of bound (^{3}H) 5αDHT (table IV). Although androgen binding activity was present in the epididymides of these very immature rabbits, it was apparent that the concentration of available binding sites for exogenous (^{3}H) 5αDHT was quite low in all three segments.

When (^{3}H) 5αDHT binding to epididymal segments of animals from the second group were examined, results shown in table IV were obtained. These data show that all three segments from individual rabbits were able to bind (^{3}H) 5αDHT. When unlabeled 5αDHT was present in the incubation mixture, the radioactive peaks were completelv eliminated, suggesting the presence of a limited number of binding sites.

In animals from the third group, (^{3}H) 5αDHT binding was only detectable in the caput (table IV). This pattern was observed in animals as young as 60 days of age and is typical of the pattern present in intact sexually mature rabbits (10).

There was a considerable overlap of the ranges of age and body weight of the animals in each binding category, indicating that there was no correlation of either of these parameters with binding activity. On the other hand, there was a significant difference ($p < 0.01$) between testes weights among the three groups, and there was an excellent correlation of testis weight with the binding pattern.

Histological Examination of Testes and Epididymides

Histological sections of the testes and the epididymides were examined and all animals were ranked in ascending order of testicular development. It was found that this order corresponded closely to the pattern of 5αDHT binding observed. In animals where there was little or no detectable binding in the three epididymal segments, the seminiferous tubules in the testis appeared as solid cords of cells containing supporting cells and either spermatogonia or spermatogonia and the first spermatocytes. The lining of the ductus epididymidis in the caput was uniform and consisted of a single row of low columnar cells with round nuclei. Stereocilia were seen only in the corpus region. In all animals in which there was binding to all three segments, the seminiferous tubules exhibited spermatogenic activity and formation of a lumen. Stereocilia were seen in the corpus and cauda. In the most mature animals of this group, stereocilia were only beginning to form in the caput. In all animals where binding was restricted to the caput, spermatogenesis had progressed from the formation of round spermatids in the most immature animals to the formation of the first elongated spermatids in the most mature ones. The diameter of the ductus epididymidis and the height of the epithelium showed a low but uniform increase throughout the epididymis with the exception of the proximal portion of the caput which was now fully differentiated and the cells had long stereocilia.

Discussion

Although 5αDHT binding can only be detected in the caput epididymidis of intact sexually mature rabbits (10), binding can be detected in all three segments of castrated animals. It is likely that the binding which is present several days after castration is due to tissue receptor and not testicular ABP. Since the entire epididymis is an androgen-dependent organ, one would anticipate the presence of receptor along the length of the duct. It is likely that the presence of endogenous androgen masks receptor binding sites in intact animals. It would appear that after removal of endogenous androgen, free sites become available for binding of exogenous (^{3}H) 5αDHT.

Following procedures that prevent testicular fluid from entering the epididymis, there is a reduction or elimination of ABP in cytosol prepared from the caput epididymidis. Similar decreases in (^{3}H) 5αDHT binding occur in caput cytosol prepared from epididymides that had been confined to the abdomen. Our observations are consistent with the hypothesis that the 5αDHT binding moiety that we described previously (10) in caput cytosol from sexually mature intact rabbits is of testicular rather than epididymal origin. Our studies with immature rabbits suggest that we are also dealing with testicular ABP in these animals. Presumably, the negligable amount of 5αDHT binding in animals of group I is due to a very small amount of ABP being produced by the testis. Animals in group II have lumina in the seminiferous tubules, thus ABP produced by the testis can readily enter the epididymis. Since the epididymal epithelium of these rabbits is relatively undifferentiated, ABP may be able to traverse the entire length of the duct and can, therefore, be detected in all segments. In animals of group III, the epithelium of the caput epididymidis has the adult appearance and presumably ABP is resorbed or inactivated in this region, preventing the detection of androgen binding in more distal segments of the epididymis.

The fate and function of the ABP and its attached androgen are not yet clear. It would appear that testicular ABP has evolved as an important factor in a mechanism whose function is to ensure a very high concentration of androgen in the proximal portion of the epididymis. Since spermatozoa undergo androgen-dependent maturational processes while passing through this portion of the epididymis, it is tempting to speculate that ABP may be involved in these processes. However, it should be noted that prevention of the entry of the rete testis fluid into the epididymis by ligation of the ductuli efferentes or by unilateral castration has no effect on sperm maturation (12), nor does it change the biochemical characteristics of the mature epididymis (*Leathem,* personal commun.). One may argue that after efferentiectomy or unilateral castration, spermatozoa tested for their fertilizing ability have already been exposed to luminal androgen present in the caput. However, if spermatozoa are prevented from moving down the epididymis and are maintained in the caput, they never mature (4, 21). Spermatozoa in epididymal segments kept in organ culture *in vitro* do not mature, but if testosterone is added to the medium, some degree of maturation is achieved (22). *In vivo* administration of exogenous testosterone permits normal sperm maturation and sperm survival in castrated and hypophysectomized animals (20).

These data suggest that the androgens responsible for the initiation of sperm maturation and maintenance of sperm viability are from the circulation and are probably not from the rete testis fluid. What could be the role of the androgens in rete testis fluid and the role of testicular ABP? After efferenctiectomy, there is a decrease in tubule diameter and epithelial height, and an increase in vacuolization of the caput epithelium (13, 20). It is possible that this region of the epididymis requires a higher level of androgen to maintain its structural integrity than do the other regions. The existence of an ABP in rabbit plasma provides another potential pool of androgens for use by epididymal tissue. In addition to the androgens brought to the epididymis by transport proteins (ABP and TeBG), a further source of androgen is provided by synthetic processes in the epididymis. Since these synthetic processes are themselves dependent on circulating androgen (15), it would seem that the most important source of androgen is the circulation.

After these considerations, it would appear that the target cell receptor that can be demonstrated in the epididymis of castrated animals is probably of primary importance in mediating the androgen-dependent processes that are required for maintaining the many complex morphological and functional aspects of the epididymis. The role of these androgen-dependent phenomena in the economy of the epididymis is not yet understood. Our current hypothesis is that one of the functions of androgens is to act on the epididymis to produce secretory products that interact with spermatozoa to bring about changes that are required for their maturation and to increase their survival. In this context, it is of interest to note that androgen-dependent protein synthesis has been demonstrated in the epididymis in organ culture by *Blaquier* (8). This hypothesis may appear simplistic, but it is difficult for us to consider that the complex androgen regulation of the epididymis is directed solely toward mediating effects in the tissue itself without affecting the enclosed spermatozoa. The testing of this hypothesis awaits future research.

Acknowledgments

The invaluable technical assistance of *Marty Sribnick* is gratefully appreciated. Supported by NIH Contract 69-2166, NIH Grants KHD-38882, HD-051541, HD-05797 and HD-00073, The Population Council, and the Ford Foundation.

References

1 *Allen, J.M. and Slater, J.J.:* A chemical and histochemical study of enzymes dephosphorylating adenosine phosphate esters in the epididymis of normal, castrated, and testosterone propionate treated castrated mice. Amer. J. Anat. *105:* 117–139 (1959).

2 *Barack, B.M.:* Transport of spermatozoa from seminiferous tubules to epididymis in the mouse: a histological and quantitative study. J. Reprod. Fertil. *16:* 35–48 (1968).

3 *Benoit, M.J.:* Recherches anatomiques, cytologiques et histophysiologiques sur les voies excrétrices du testicule, chez les mammifères. Arch. Anat.-Histol. Embryol. *5:* 173–412 (1926).

4 *Bedford, J.M.:* Effects of duct ligation on the fertilizing ability of spermatozoa from different regions of the rabbit epididymis. J. exp. Zool. *166:* 271–282 (1967).

5 *Blaquier, J.A.; Cameo, M.S., and Charreau, E.H.:* Comparative uptake of androstene-diol, testosterone, and dihydrotestosterone by tissues of the male rat. J. Steroid Bio-chem. *1:* 327–334 (1970).

6 *Blaquier, J.A.:* Selective uptake and metabolism of androgens by rat epididymis. The presence of a cytoplasmic receptor. Biochem. biophys. Res. Commun. *45:* 1076–1082 (1971).

7 *Blaquier, J.A. and Calandra, R.S.:* Intranuclear receptor for androgens in rat epidid-ymis. Endocrinology *93:* 51–60 (1973).

8 *Blaquier, J.A.:* An *in vitro* action of androgens on protein synthesis by epididymal tubules maintained in organ culture. Biochem. biophys. Res. Commun. *52:* 1177–1183 (1973).

9 *Cavazos, L.F.:* Effects of testosterone propionate on histochemical reactions of epi-thelium of rat ductus epididymidis. Anat. Rec. *132:* 209–227 (1958).

10 *Danzo, B.J.; Orgebin-Crist, M.-C., and Toft, D.O.:* Characterization of a cytoplasmic receptor for 5α-dihydrotestosterone in the caput epididymidis of intact rabbits. Endo-crinology *92:* 310–317 (1973).

11 *Danzo, B.J. and Orgebin-Crist, M.-C.:* Androgen regulation of epididymal function; in *Hafez and Thibault* Transport, Survie et Pouvoir fécondant des Spermatozoïdes chez les Vertébrés, pp. 61–87 (INSERM, Paris 1973).

12 *Dyson, A.L.M.B. and Orgebin-Crist, M.-C.:* Effect of hypophysectomy, castration, and androgen replacement upon the fertilizing ability of rat epididymal spermatozoa. Endocrinology *93:* 391–403 (1973).

13 *Dyson, A.L.M.B.:* The maturation of rat spermatozoa; dissertation, London (1972).

14 *French, F.S. and Ritzén, E.M.:* A high-affinity androgen binding protein (ABP) in rat testis: evidence for secretion into efferent duct fluid and absorption by epididymis. Endocrinology *93:* 88–95 (1973).

15 *Hamilton, D.W.; Jones, A.L., and Fawcett, D.W.:* Cholesterol biosynthesis in the mouse epididymis and ductus deferens: a biochemical and morphological study. Biol. Reprod. *1:* 167–184 (1969).

16 *Hansson, V. and Tveter, K.J.:* Uptake and binding *in vivo* of ^{3}H-labeled androgen in the rat epididymis and ductus deferens. Acta endocrin., Kbh. *66:* 745–755 (1971).

17 *Hansson, V.; Djøseland, O.; Reusch, E.; Attramadal, A., and Torgessen, O.:* An andro-gen binding protein in the testis cytosol fraction of adult rats. Comparison with the androgen binding protein in the epididymis. Steroids *21:* 457–474 (1973).

18 *Lubicz-Nawrocki, C.M. and Glover, T.D.:* The influence of the testis on the survival of spermatozoa in the epididymis of the golden hamster *(Mesocricetus auratus).* J. Reprod. Fertil. *34:* 315–329 (1973).

19 *O'Malley, B.W.; Rosenfeld, G.C.; Comstock, J.P., and Means, A.R.:* Induction of specif-ic translatable messenger RNAs by oestrogen and protesterone. 5th Karolinska Symp. Research Methods in Reproductive Endocrinology, pp. 381–395 (1972).

20 *Orgebin-Crist, M.-C.; Davies, J., and Tichenor, P.:* Maturation of spermatozoa in the rabbit epididymis: effect of hypophysectomy and castration; in *Segal, Crozier, Corf-man and Condliffe* Regulation of mammalian reproduction, pp. 189–203 (Ch.C. Thomas, Springfield 1972).

21 *Orgebin-Crist, M.-C.:* Sperm maturation in the rabbit epididymis. Nature, Lond. *216:* 816–818 (1967).

22 *Orgebin-Crist, M.-C. and Tichenor, P.:* Effect of testosterone on sperm maturation *in vitro.* Nature, Lond. *245:* 328–329 (1973).

23 *Raspé, G.* (ed.): Schering workshop on hormone receptors. Adv. Biosci., vol. 7 (Pergamon Press, Oxford 1971).
24 *Ritzén, E.M; Nayfeh, S.N.; French, F.S., and Dobbins, M.C.:* Demonstration of androgen-binding components in rat epididymis cytosol and comparison with binding components in prostate and other tissues. Endocrinology *89:* 143–151 (1971).
25 *Ritzén, E.M.; Dobbins, M.C.; Tindall, D.J.; French, F.S., and Nayfeh, S.N.:* Characterization of an androgen binding protein (ABP) in rat testis and epididymis. Steroids *21:* 593–607 (1973).
26 *Setchell, B.P.:* Testicular fluids; in *Johnson, Gomes and Vandemark* The testis, vol. 1, p. 196 (Academic Press, New York 1970).
27 *Steggles, A.W.; Spelsberg, T.C.; Glasser, S.R., and O'Malley, B.W.:* Soluble complexes between steroid hormones and target-tissue receptors bind specifically to target-tissue chromatin. Proc. nat. Acad. Sci. Wash. *68:* 1479–1482 (1971).
28 *Tindall, D.J.; French, F.S., and Nayfeh, S.N.:* Androgen uptake and binding in rat epididymal nuclei *in vivo.* Biochem. biophys. Res. Commun. *49:* 1391–1397 (1972).
29 *White, W.E.:* The effect of hypophysectomy on the survival of spermatozoa in the male rat. Anat. Rec. *54:* 253–273 (1932).

Authors' address: Dr. *B.J. Danzo,* Prof. *Marie-Claire Orgebin-Crist* and Mrs. *Barbara C. Eller,* Department of Obstetrics and Gynecology, Vanderbilt University School of Medicine, *Nashville, TN 37232* (USA)

The Biology of Spermatozoa
INSERM Int. Symp., Nouzilly 1973, pp. 28–35 (Karger, Basel 1975)

Transport and Maturation of Epididymal Spermatozoa in Domestic Animals[1]

Claude Esnault, Michel Courot and Robert Ortavant
INRA, Station de Physiologie de la Reproduction, Nouzilly

During passage through the epididymis, the spermatozoa of most species undergo a process of maturation by which they acquire progressive motility and the capacity for fertilization. However, spermatozoa retain these properties for only a limited time, and therefore it is important to determine the normal duration of their passage through the epididymis.

The period of time separating appearance of labelled spermatozoa in the first part of the head of the epididymis and their reappearance in the ejaculate measures the time of transit along the male genital tract; this has been found to be from 8 to 15 days according to the species (23). The delay between injection of radioactive precursor and release into the lumen of seminiferous tubules of the first labelled spermatozoon is constant (22). Frequency of sperm collection does not affect the time of appearance of labelled spermatozoa in the head of the epididymis, but modifies that of labelled spermatozoa in the ejaculate (19). Furthermore, all gametes do not pass along the epididymis at the same rate, so there is a mixture of spermatozoa of different ages in the tail of the epididymis.

Little work has been done on the acquisition of fertilizing capacity of spermatozoa in domestic animals. To our knowledge, the literature on this subject does not contain any reports of fertile artificial insemination of testicular spermatozoa or spermatozoa taken from the head of the epididymis in large animals. On the other hand, spermatozoa taken from the tail of the epididymis of the ram (9) or the bull are usually capable of successful fertilization (15).

In a detailed study of 40 bulls, *Branton* (5) observed no motility in spermatozoa taken from the testis or in most spermatozoa taken from the head of the

1 The authors are indebted to Dr. *B.P. Setchell* for English translation and to Mrs. *A. Daveau*, Mr. *J.C. Nicolle* and Mr. *C. Cornu* for their skillful assistance. A part of this work was supported by DGRST contracts No. 71-7-3156 and 72-7-0098.

epididymis. However, 80 % of samples from the tail of the epididymis contained spermatozoa which were normally motile.

In the ram, *Salamon* (24) observed an increased proportion of motile spermatozoa (55.8 %) in samples from the head of the epididymis diluted in physiological serum. This proportion increased as the spermatozoa progressed along the epididymis.

All authors recognize a morphological evolution of spermatozoa during the course of epididymal transit in domestic animals. The presence of cytoplasmic droplets is more frequently seen in spermatozoa from the head than in those from the tail of the epididymis. These move from the base of the head to the distal part of the middle piece during the progression of spermatozoa along the epididymal canal. Membrane permeability, revealed by staining with eosin, increases to a maximum in the body of the epididymis, suggesting that this part plays a particular role (14, 21).

Among the constituents of spermatozoa, we will consider the evolution of deoxyribonucleic acid (DNA) and basic nucleic proteins during spermatozoon maturation.

Deoxyribonucleic Acid

From evidence obtained by microspectrophotometry, it has been shown that the level of DNA-UV in spermatozoa does not change during the course of epididymal transit. On the contrary, modifications of the DNA level have been observed when measured by Feulgen staining (DNA-Feulgen). *Gledhill et al.* (13) have produced evidence that in the bull an important fall in the DNA-Feulgen of spermatids occurs at the end of spermiogenesis, this decrease continuing, in the course of epididymal transit, to spermatozoa in the ejaculate. *Bouters et al.* (3) and *Bouters and Vandeplassche* (4) have shown an identical phenomenon in rabbit and bull, spermatozoa in the ampullae and the ejaculate having appreciably lower values for DNA-Feulgen than those of the head and tail of the epididymis. At the time of release of spermatozoa into the lumen of the seminiferous tubules of the ram, *Esnault* (10) observed a sudden decrease in the level of DNA-Feulgen in comparison to that of the elongated spermatids, while the quantity of DNA-UV did not show any variation (table I). During passage along the epididymis the spermatozoa show DNA-Feulgen values lower than those of testicular spermatozoa, the lowest level being observed in the spermatozoa from the ejaculate. This phenomenon cannot be explained by differences in sensitivity of the HCl hydrolysis involved in the Feulgen reaction, since the optimum time for hydrolysis is the same for each of the different cellular categories (10).

It is likely that the Feulgen reaction in the spermatozoa is affected by factors concerned with the qualitative and quantitative variations of the basic proteins of the nucleus, from the time of condensation of the chromatin and through the different stages of maturation of the male gamete.

Calvin and Bedford (7) have shown that treatment with a mixture of a surface active agent (sodium dodecyl sulphate, SDS) and a specific reducer of the disulphide linkages (dithiothreitol, DTT) provokes a decondensation which is more apparent in the spermatozoa of the head than in those of the tail of the epididymis, or in those ejaculated. This suggests a progressive increase of SS-linkages in the spermatozoal nucleus during its passage through the epididymis. Spermatozoa treated with DTT also retain 3–5 times more actinomycin-D than the controls; this treatment increases their capacity of binding tritiated-AMD to a level close to that of immature spermatozoa (8). It is possible to obtain experimental reactivation of the Feulgen reaction by treating spermatozoa with dithiothreitol (10). After this treatment epididymal spermatozoa or ejaculated

Table I. Microspectrophotometric measurements on ram germinal cells during spermiogenesis and transit through the epididymis

Cells	Feulgen-DNA, 560 nm	UV-DNA, 265 nm
Spermatids, stages 6 and 7	7.54 ± 0.043	5.46 ± 0.044
Testicular spermatozoa	4.85 ± 0.036	5.38 ± 0.035
Spermatozoa from the caput epididymidis	3.44 ± 0.022	5.55 ± 0.061
Spermatozoa from the corpus epididymidis	2.97 ± 0.017	5.56 ± 0.053
Spermatozoa from the cauda epididymidis	2.73 ± 0.012	5.65 ± 0.034
Ejaculated spermatozoa	2.19 ± 0.017	5.60 ± 0.031

Results expressed as mean total extinction values ± SE in arbitrary units for 60 cells (UV-DNA) and 120–240 cells (Feulgen-DNA) for each class.

Table II. Effect of dithiothreitol on total extinction at 560 nm of ram spermatozoa stained by the Feulgen reaction

Spermatozoa from the	Control	Dithiothreitol 15 min	
		class 1	class 2
Caput	3.85 ± 0.029	3.81 ± 0.035	7.33 ± 0.028
Corpus	3.76 ± 0.026	3.65 ± 0.026	7.49 ± 0.028
Cauda	3.45 ± 0.020	3.32 ± 0.049	7.16 ± 0.021
Ejaculate	3.10 ± 0.026	3.31 ± 0.021	7.94 ± 0.034

Results expressed as mean total extinction values ± SE in arbitrary units for 90 cells of each class.

spermatozoa can be divided into two classes according to their staining (table II, fig. 1 d): (1) spermatozoa in which the nucleus is like that of the controls; (2) spermatozoa with the nucleus very heavily stained, lightly granulated, the total extinction of which is identical to that of elongating spermatid nuclei.

The latter class represents 13–94 % of spermatozoa according to the sample. The nuclear surface of the spermatozoa is equal to or very slightly less than that of the controls.

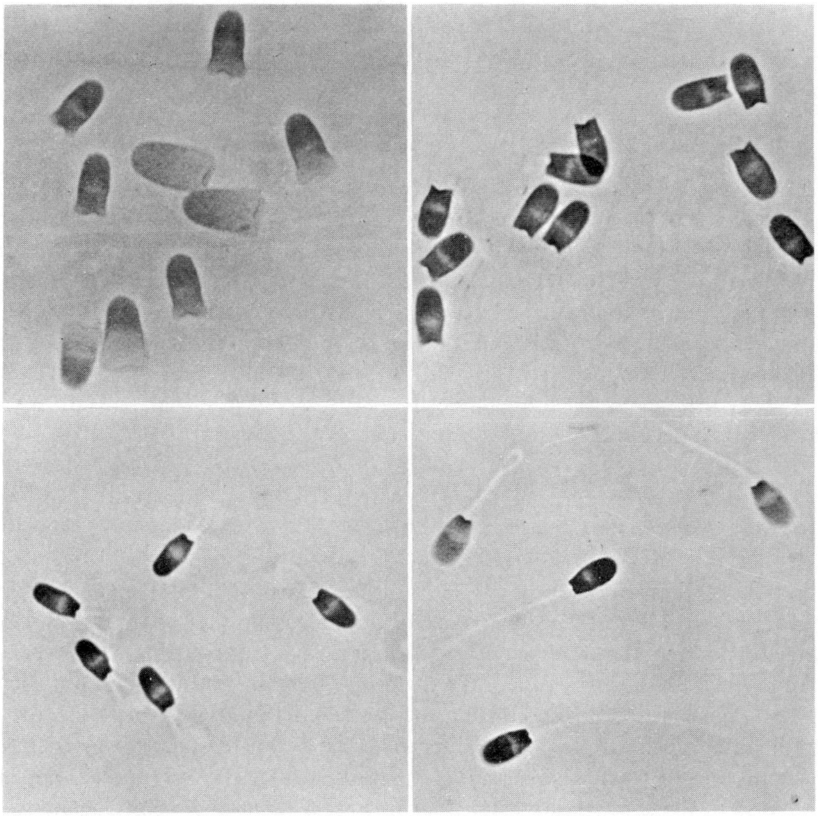

a, b

c, d

Fig. 1. (a) Ram spermatozoa from the caput epididymidis exposed to 1 % sodium dodecyl sulfate (SDS) – 0.05 *M* borate buffer containing 2×10^{-3} *M* dithiothreitol (DTT) for 30 min. Feulgen staining. Note the gross swelling of the nuclei and the absence of tail components. (b) Ram spermatozoa from the corpus epididymidis (same treatment). (c) Ram spermatozoa from the cauda epididymidis (same treatment). (d) Ejaculated ram spermatozoa exposed to DTT-borate buffer without SDS. Two nuclei are very densely stained by the Feulgen reaction.

Table III. Ram spermatozoa: effect of a mixture of dithiothreitol (DTT) and sodium dodecyl sulfate (SDS) on Feulgen-DNA (total extinction) and area (μm^2)

Treatment		Epididymal spermatozoa from		
		caput	corpus	cauda
Control	Total extinction	3.70 ± 0.065	3.81 ± 0.058	3.56 ± 0.029
	Area	34.80 ± 0.326	32.87 ± 0.261	31.93 ± 0.283
SDS + DTT, 30 min	Total extinction	4.38 ± 0.055	4.45 ± 0.061	4.99 ± 0.031
	Area	95.80 ± 1.873	47.13 ± 1.734	27.20 ± 0.587

Results expressed as mean total extinction values ± SE in arbitrary units for 30 cells of each class.

When a detergent (SDS) is added to the dithiothreitol (DTT), a marked and very rapid swelling is observed in spermatozoa from the head of the epididymis, less in those from the body, and not at all in ejaculated spermatozoa or those from the tail of the epididymis. Total extinction by microspectrophotometry of nuclei stained by the Feulgen reaction after this treatment is intermediate between those of the controls and those of the spermatozoa treated by DTT alone (table III, fig. 1 a, b, c).

This reactivation is comparable to that observed in ejaculated ram spermatozoa which have spent some time in the uterus of the ewe.

Basic Nuclear Proteins

The studies of *Lavon et al.* (17) suggest a loss of total protein during the course of epididymal transit, in spite of an increase in the proportion of certain amino acids, such as arginine and cystine, in the nuclear proteins. The increase in basicity of nuclear proteins and the quantity of arginine in spermatids at the end of spermiogenesis continues during the passage of the spermatozoa through the epididymis, although to a lesser degree (13). *Loir* (18) has shown that these proteins in the ram, which appear in the nuclei of old spermatids, are more stable than those synthesized in other phases of meiosis or spermiogenesis, and that they do not take part in turnover, at least until the end of epididymal transit.

In the ram (11), a quantitative cytochemical study has revealed that the level of arginine increases quite rapidly in the nuclei of spermatids after the phase of elongation. A maximum is reached in testicular spermatozoa and does not change to any extent during epididymal transit. To quantitate staining by alkaline fast green (1) – in order to follow the evolution of histones – we studied different times of hydrolysis using 8 % trichloracetic acid at 90 °C. This concentration avoids an eventual elimination of the lysine-rich proteins (16). Results show that the duration of hydrolysis, necessary to completely extract DNA from the epididymal spermatozoa, increases progressively from the head to the tail of the epididymis. On the other hand, the quantity of proteins rich in lysine and arginine is quite small in round spermatids, but increases slightly during stages of elongating spermatids, and very sharply during the following stages of spermiogenesis, to a maximum in testicular spermatozoa. During epididymal transit a decrease in the stainability of histones can be observed. This begins at the level of the head of the epididymis and becomes more marked in the spermatozoa found in the body. The spermatozoa in the tail of the epididymis and ejaculated spermatozoa show values for extinction which are approximately equivalent to those of testicular spermatozoa (fig. 2).

The last synthesis of histones, having taken place for lysine in stage 3 and for arginine in stage 5 of the seminiferous epithelium cycle (18), shows that the total quantity of basic nucleoproteins in spermatozoa cannot vary during the course of passage through the epididymis. Therefore, the apparent transient decrease in histone stainability must be due to some physicochemical modifications to the nucleus of the spermatozoa in the head and body of the epididymis which make them particularly fragile. This would induce histone extraction during the course of the treatment or prevent stoichiometry with fast green. Numerous studies have shown an evolution of the characteristics of the male gamete in the body of the epididymis: increase of the permeability of eosin (14, 21); acquisition of fertilizing ability (20); development of an attraction of colloidal iron to the plasma membrane (2); appearance of dense bodies in the

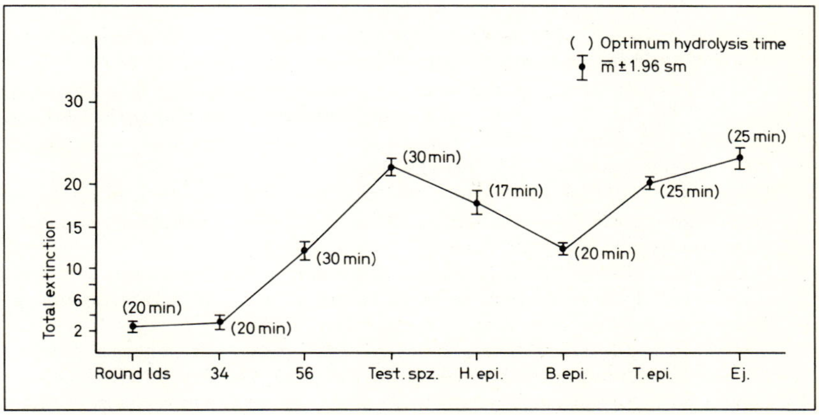

Fig. 2. Ram – basic proteins (fast green stain, pH 8.2) 640 nm.

acrosome (12); interaction between zinc and disulphide groups which is more important in spermatozoa in the tail of the epididymis than in those of the head (6).

Thus, the analysis of basic proteins in the germinal cells of domestic animals reveals an increase at the end of spermiogenesis, and qualitative or quantitative variations among certain of them during epididymal transit. Probably histone binding of a germinal type, rich in arginine and cystine, with deoxyribonucleic acid, confers on the spermatozoa their great resistance when confronted with most chemical or enzymatic denaturing agents.

References

1 *Alfert, M. and Geschwind, I.I.:* A selective staining method for the basic proteins of cell nuclei. Proc. nat. Acad. Sci., Wash. *39:* 991–999 (1953).

2 *Bedford, J.M.; Calvin, H., and Cooper, G.W.:* The maturation of spermatozoa in the human epididymis. J. Reprod. Fertil. Suppl. *18:* 199–213 (1973).

3 *Bouters, R.; Esnault, C.; Salisbury, G.W., and Ortavant, R.:* Discrepancies in analysis of deoxyribonucleic acid in rabbit spermatozoa, involving Feulgen staining (Feulgen-DNA) and ultraviolet light absorption (UV-DNA) measurements. J. Reprod. Fertil. *14:* 355–363 (1967).

4 *Bouters, R. und Vandeplassche, M.:* Vergleichende mikrospectrophotometrische Bestimmungen von Feulgen-DNS an epididymären, ampullären, ejakulierten und durch Massage gewonnenen Spermien von Bullen. Zuchthygiene *3:* 1–5 (1968).

5 *Branton, C.:* Physiology and morphology of spermatozoa from different levels of the reproductive tract of the bull; M. Sci. thesis, Ithaca (1946).

6 *Calvin, A.:* Keratenoid proteins in the heads and tails of mammalian spermatozoa. Biol. J. Linn. Soc. Suppl. (in press).

7 *Calvin, H.I. and Bedford, J.M.:* Formation of disulphide bonds in the nucleus and accessory structures of mammalian spermatozoa during maturation in the epididymis. J. Reprod. Fertil. Suppl. *13:* 65–75 (1971).

8 *Calvin, H.I.; Yu, C.C., and Bedford, J.M.:* Influence of disulfide bonding on the binding of actinomycin D to mammalian sperm chromatin. Anat. Rec. *175:* 283 (1973).

9 *Dott, H.M.:* The fertility of ram semen treated in the continuous flow dialysis apparatus. Proc. 4th Int. Congr. of Animal Reproduction, vol. 4, pp. 873–876 (1961).

10 *Esnault, C.:* Reactivation of the Feulgen reaction of ram spermatozoa by dithiothreitol. J. Reprod. Fertil. *32:* 153–157 (1973).

11 *Esnault, C.:* Evolution de l'ADN et des protéines nucléaires basiques au cours de la maturation des cellules germinales du bélier. Etude microspectrophotométrique. Coll. Int. d'Histochimie, Tours 1973.

12 *Fléchon, J.E.:* Etude ultrastructurale et cytochimique des constituants de la tête et de la surface du spermatozoïde de lapin. Modifications au cours de la maturation et de la capacitation *in utero;* thèse doct. 3e cycle, Paris (1973).

13 *Gledhill, B.L.; Gledhill, M.P.; Rigler, R., jr., and Ringertz, N.R.:* Changes in deoxyribonucleoprotein during spermiogenesis in the bull. Exp. Cell Res. *41:* 652–665 (1966).

14 *Glover, T.D.:* Disintegrated spermatozoa from the epididymis. Nature, Lond. *190:* 185–186 (1961).

15 *Igboeli, G. and Foote, R.H.:* Maturation changes in bull epididymal spermatozoa. J. Dairy Sci. *51:* 1703–1705 (1968).
16 *Kinkade, J.M., jr. and Cole, R.C.:* The resolution of four lysine-rich histones derived from calf thymus. J. biol. Chem. *241:* 5790–5797 (1966).
17 *Lavon, U.; Volcani, R., and Danon, D.:* The proteins of bovine spermatozoa from the caput and cauda epididymis. J. Reprod. Fertil. *24:* 219–232 (1971).
18 *Loir, M.:* Métabolisme de l'acide ribonucléique et des protéines dans les spermatocytes et les spermatides du bélier *(Ovis aries).* II. Variations de l'incorporation et devenir de la ^{3}H-lysine, de la ^{3}H-arginine et de la ^{35}S-cystine. Ann. Biol. anim. *12:* 411–429 (1972).
19 *Orgebin-Crist, M.-C.:* Recherches expérimentales sur la durée de passage des spermatozoïdes dans l'épididyme du taureau. Ann. Biol. anim. *2:* 51–108 (1962).
20 *Orgebin-Crist, M.-C.:* Studies on the function of the epididymis. Biol. Reprod. Suppl. *1:* 155–175 (1969).
21 *Ortavant, R.:* Existence d'une phase critique dans la maturation épididymaire des spermatozoïdes de bélier et de taureau. C.R. Soc. Biol. *147:* 1552–1556 (1953).
22 *Ortavant, R.:* Autoradiographie des cellules germinales du testicule de bélier. Durée des phénomènes spermatogénétiques. Arch. Anat. micr. Morph. exp. *45:* 1–10 (1956).
23 *Ortavant, R.; Orgebin-Crist, M.-C. et Singh, G.:* Etude comparative de la durée des phénomènes spermatogénétiques chez les animaux domestiques. Symp. on The Use of Radioisotopes in Animal Biology and the Medical Sciences, Mexico, pp. 321–327 (Academic Press, New York 1961).
24 *Salamon, S.:* Viability and morphology of ram spermatozoa in the ligated epididymis. Austr. J. Biol. Sci. *21:* 769–780 (1968).

Authors' address: *C. Esnault, M. Courot and R. Ortavant,* Institut National de la Recherche Agronomique, Laboratoire de Physiologie de Reproduction, *Nouzilly, F–37380 Monnaie* (France)

The Biology of Spermatozoa
INSERM Int. Symp., Nouzilly 1973, pp. 36–45 (Karger, Basel 1975)

Ultrastructural and Cytochemical Modifications of Rabbit Spermatozoa during Epididymal Transport

Jacques-E. Fléchon

INRA, Station de Physiologie animale, Jouy-en-Josas

Introduction

During epididymal transport, the acquisition of fertilizing ability is accompanied by some well-known morphological and physiological changes in the spermatozoa such as modification of acrosomal shape, loss of cytoplasmic droplet and progressive increase of motility (18).

In the rabbit, sperm maturation has been the subject of numerous studies and the epididymal segment where spermatozoa begin to be fertilizing is known (17).

We have used this species to follow the ultrastructural and cytochemical evolution of spermatozoa during migration in the epididymis. Our study is devoted to the acrosome and cell coat in view of the evident role these parts of the male gamete play in fertilization.

Materials and Methods

Epididymal material or ejaculated sperm were obtained from the same rabbit bucks of known fertility. The three levels of the epididymis used were the segments 3, 7 and 8 of *Nicander*'s classification (14). They will be called the caput, corpus and cauda epididymis.

Pieces of epididymal duct and epididymal or ejaculated spermatozoa were fixed with glutaraldehyde generally without subsequent osmication, and embedded in epoxy resins or glycolmethacrylate (GMA).

General staining of thin sections was achieved by means of uranyl acetate and lead citrate, and specific staining of glycoproteins was done on thin sections in GMA with phosphotungstic acid in normal hydrochloric acid (PTA/HCl) at pH 0.3.

Positively charged colloidal iron hydroxide (CIH) was used at low pH en bloc or on thin GMA sections for the detection of acidic groups.

Enzymatic extractions on glutaraldehyde-fixed spermatozoa were performed with pronase and *Vibrio cholerae* neuraminidase before embedding, or on thin sections.

A complete account of technical details was given previously (9).

Results

Structural and Cytochemical Modifications of Acrosome during Epididymal Transport

Appearance of Crystalline Cores

In the caput epididymis, the marginal thickening of the acrosome is relatively thin and elongated and the equatorial segment is sometimes thicker at its posterior end. The acrosomal content is generally homogeneous, showing principally a subsurface lamina in the anterior segment (fig. 1).

In the corpus and cauda epididymis and ejaculated spermatozoa the equatorial segment is very thin and the marginal thickening is shorter and more dilated. The acrosomal material is heterogeneous, particularly in the marginal thickening where an electron-dense core is located on both faces of the sperm

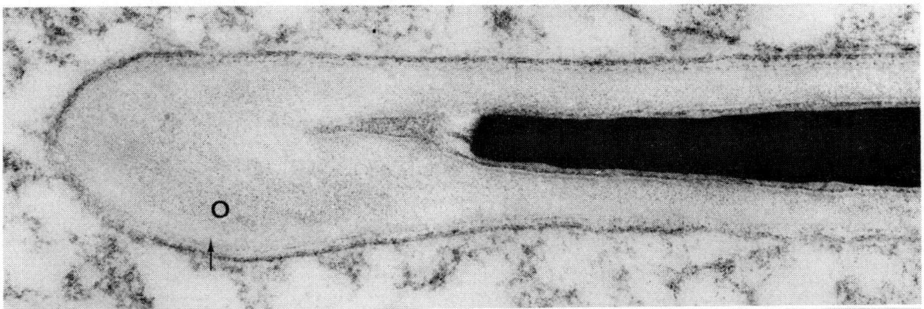

1

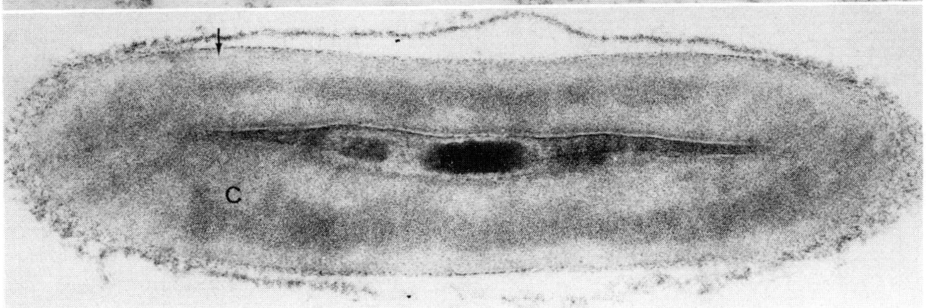

2

Fig. 1. Sagittal section through the anterior part of a sperm head in the caput epididymis. The marginal thickening is long and flat. Its content is rather homogeneous, showing a slight opacity (o) announcing the dense cores (?) and a superficial lamina (→). Glutaraldehyde, Epon, U and Pb. × 35,000.

Fig. 2. Transverse section through the anterior part of a sperm head in the caput epididymis. The rounded marginal thickening contains a superficial lamina (→) and crystalline cores (C); periodicity of 70 Å. Glutaraldehyde, Epon, U and Pb. × 35,000.

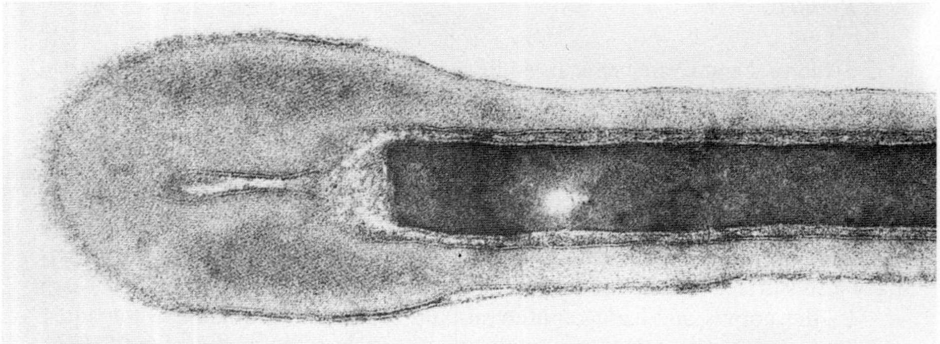

3

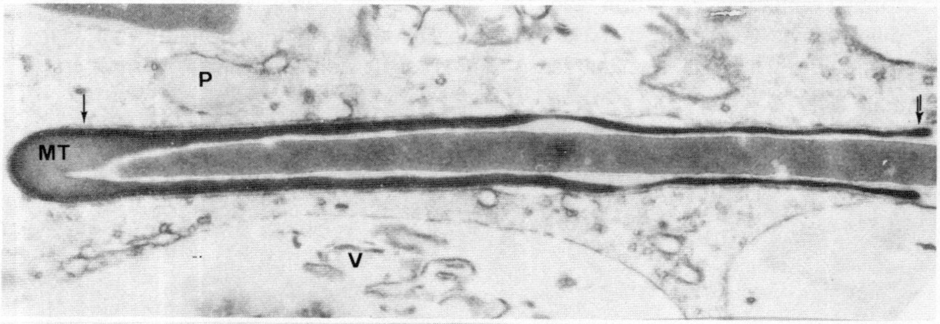

4

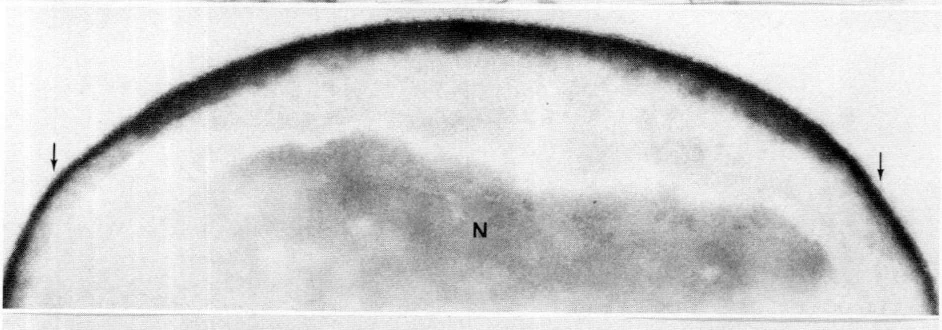

5

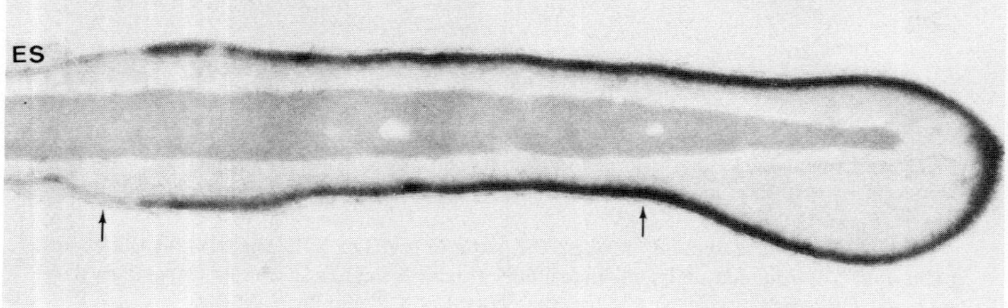

6

head. As the marginal thickening itself, these two bodies are horseshoe-shaped and show similar dorso-ventral dissymmetry.

At high magnification and on thin sections in epoxy resins the dense cores appear as alternate clear and dark lines at intervals of 60–90 Å, oriented in one or two directions, depending on the plane of sectioning (fig. 2, 3).

Cytochemical Reactions on Acrosome

In the spermatozoa of the caput epididymis (fig. 4), PTA/HCl stains the entire acrosomal content (except the inner mass of the marginal thickening). Starting from the corpus epididymis, the stainability of the equatorial segment is lost, but that of the anterior segment is not modified noticeably along the epididymal duct (fig. 5–7). The dense cores appear in the internal region of the marginal thickening which is not stained by PTA/HCl.

The content of the marginal thickening is particularly sensitive to a short pronase treatment of thin sections in GMA. The whole acrosomal substance is digested after a long incubation of thin sections in epoxy resins with the enzyme.

Cytochemical Modifications of Cell Coat during and after Epididymal Transport

Evolutive Distribution of Surface Glycoproteins

A continuous PTA/HCl stained cell coat appears on the whole surface of spermatozoa in the lumen of the caput epididymis which also contains stained material (fig. 4).

In the corpus epididymis the staining begins to be discontinuous on the

Fig. 3. Oblique section through the anterior part of a sperm head after ejaculation and *in vitro* storage. Crystalline cores in the marginal thickening (periodicity 70–90 Å). Glutaraldehyde, OsO_4, Epon, U and Pb. × 49,000.

Fig. 4. Sagittal section through the acrosomal region of a sperm head in the corpus epididymis. Selective staining of glycoproteins in the equatorial segment–slightly swollen at the posterior end ⇒ – and a superficial layer of the anterior segment particularly visible at the surface of the flat marginal thickening (MT). The cell coat is stained on the head (→), cytoplasmic droplets containing vesicles (V) and a principal piece (P). Glutaraldehyde, GMA, PTA/HCl. × 15,000.

Fig. 5. Flat section of the anterior part of a sperm head after ejaculation. Staining of glycoproteins in a superficial layer of the marginal thickening and on the thin cell coat (→). N = Nucleus. Glutaraldehyde, GMA, PTA/HCl. × 22,500.

Fig. 6. Sagittal section of the anterior part of a sperm head after ejaculation. Staining of glycoproteins in a superficial layer of the anterior segment and in the thin cell coat (→). The equatorial segment (ES) is thin and unstained. Glutaraldehyde, GMA, PTA/HCl. × 22,500.

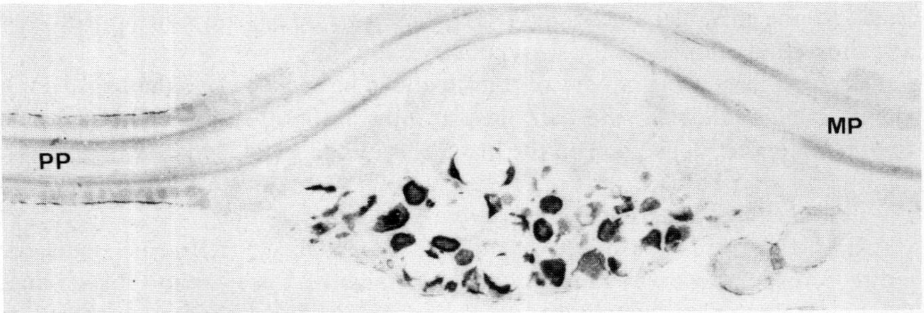

7

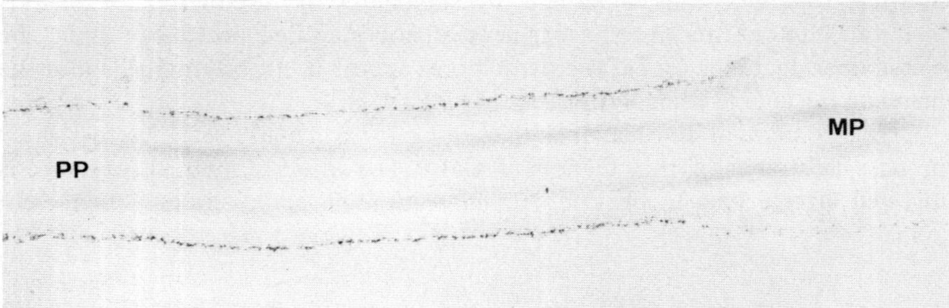

8

Fig. 7. Longitudinal section of a sperm flagellum in the cauda epididymis. Staining of a thick and discontinuous glycoprotein cell coat on the principal piece (PP) and also on transverse sections, but not on middle piece (MP). Cytoplasmic droplet at the end of migration, containing stained vesicles (V). Glutaraldehyde, GMA, PTA/HCl. × 10,000.

Fig. 8. Longitudinal section of a sperm flagellum after ejaculation. Iron particles in high density on the principal piece (PP) and very few on the middle piece (MP), visualizing distribution of the negative surface charges. Glutaraldehyde, CIH pH 1.7, GMA. × 15,000.

middle piece. On spermatozoa of the cauda epididymis and after ejaculation, the cell coat is thin on the head (fig. 5, 6), barely visible on the middle piece, and thickest on the principal piece (fig. 7) where some discontinuities are visible in ejaculated sperm.

No deposition of CIH occurs on the spermatozoa of the caput epididymis, except on some principal pieces. In lower segments of the epididymis, all principal pieces show a heavy metallic deposit and some principal pieces a slight one as in ejaculated spermatozoa (fig. 8). In no case CIH reacts with the head surface. Results on the cell coat of the flagellum are summarized in figure 9.

Characterization of Surface Glycoproteins

Effects of enzymatic extraction of spermatozoa before PTA/HCl or CIH treatments and of pH variations of the latter are summarized in table I.

Washing of ejaculated spermatozoa does not induce any visible modification of the cell coat, but seems to enhance CIH deposition on the principal piece.

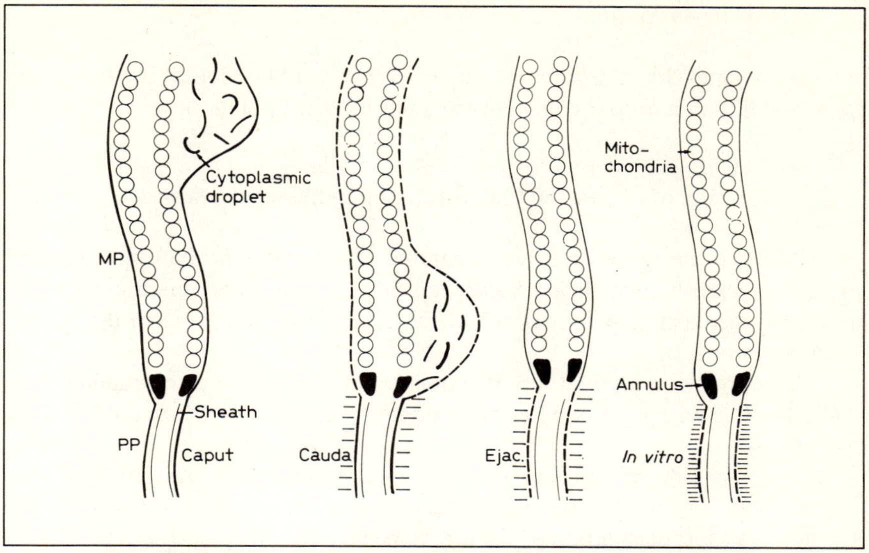

Fig. 9. Evolution of flagellum cell coat during and after epididymal transport in the rabbit. —— = Material stained with PTA/HCl; - - - = discontinuous staining; ≡ = heavy condensation of negative charges (CIH).

Table I. Composition of cell coat on the principal piece of ejaculated rabbit spermatozoa

Specific reagents	Enzymatic treatments before embedding	Reaction	Components
PTA/HCl on thin GMA sections	control	+	glycoproteins
	pronase	–	
CIH, pH ⩾ 1.7	control	+	carboxylic acids carried by (glyco) proteins
	pronase	<	
CIH, pH = 2.5	control	+	sialoproteins?
	neuraminidase on thin GMA sections	< or =	
CIH, pH 1	none	+	sulfates and/or phosphates

+, Positive; –, negative; <, less than control; =, equal to control.

Comments on Results

As an extensive discussion of our work has been previously published (9), we would like to limit our comments to a few essential points.

Modification of Acrosomal Content during Epididymal Transport

Parallel to morphological modifications of the marginal thickening (5) and equatorial segment, which are less spectacular than in other species (8), structural and chemical transformations of the acrosomal content occur during epididymal transport.

The principal structural modification is the appearance of crystalline cores inside the marginal thickening corresponding to dense cores described by a few authors (9).

The crystalline structure of these intraacrosomal bodies was observed after different fixation procedures and freeze-etching. Periodic structures were also described in superficial layers of the acrosome (9, 10).

Concerning the chemical composition of the acrosome, it should be remembered that it is PAS-positive and that extracts were found to contain glycolipoproteins (9). Whereas the whole acrosomal substance is pronase digestible, only a (large) part of this material is stainable by PTA/HCl, known to be selective for polysaccharides and glycoproteins in the conditions used (11). Thus, we have localized two components, glycoproteins and proteins, at the ultrastructural level. The latter show amorphous and crystalline regions in mature spermatozoa.

Below the caput epididymis, the loss of PTA/HCl stainability of the equatorial segment corresponds not only to a decrease in thickness but also to a migration or chemical modification of the content. Our observations confirm the low PAS stainability, if any, of the equatorial segment of mature spermatozoa (15).

Modifications of Cell Coat during and after Epididymal Transport

Our study shows that spermatozoa possess a glycoprotein cell coat, as do somatic cells, which is regularly distributed and generally deprived of free acidic groups in the caput epididymis.

Consecutive modifications of the cell coat led progressively to the state of ejaculated spermatozoa where there are regional differences in thickness of glycoproteins and abundance of free acidic groups. Unequal distribution of particles in the plasma membrane of ejaculated spermatozoa was also observed (10).

Concentration of free acidic groups on the principal piece of rabbit spermatozoa explains the higher negative surface charge of the tail and the orientation of spermatozoa in an electric field (2). Such a differentiation is exceptional in isolated cells and could be related to different functions of the various parts of the male gamete. However, the cell coat may be homogeneous in spermatozoa of other species, e.g. bull (12).

The latter study and figure 9 show that the cell surface is constituted of a mosaic of radicals. Like other authors (20), we were unable to ascertain the presence of free sialic acids on the cell coat of rabbit spermatozoa. The nature of surface sugars has been investigated by means of lectins (16).

Relationship of Observed Sperm Modifications to Epididymal Maturation

The acrosome contains various enzymes necessary for the penetration of spermatozoa through the egg envelopes (21). We have put forward the hypothesis that our ultrastructural localization of proteins and glycoproteins in the acrosome gives a picture of the distribution of, e.g. acrosine, which is probably a protein-like trypsin, and hydrolases, like hyaluronidase, which are glycoproteins (9).

In fact, the inner part of the anterior segment, particularly in the marginal thickening, appears proteinaceous and probably corresponds to the localization of acrosine (6). Moreover, the anterior segment, which alone contains glycoproteins, reacts positively to an immunofluorescent test for the localization of hyaluronidase (unpublished results).

Observed redistribution and structure modification of the acrosomal (enzymatic) content could correspond to part of the maturation process.

It seems that observed surface modifications during epididymal transport reflect true changes of the plasma membrane also expressed in ultrastructure (5) and electrophoretic mobility (2).

The origin of the cell coat of spermatozoa is not known. It seems that the spermatozoa do not possess an entirely autonomous program of differentiation, since charges of mature cells are not acquired by spermatozoa maintained in the caput epididymis by ligation (4).

As many sperm surface antigens are produced in accessory glands and in the epididymis (1, 19), modifications of the sperm cell coat could be at least partially explained by adsorption and/or loss of (glyco)proteins in the male genital tract. We observed that storage of spermatozoa *in vitro* can increase the number of free acidic groups.

Whatever the determinism of cell coat modification is, these surface changes could contribute to the phenomenon of maturation in multiple ways: (1) alteration of the adhesivity of spermatozoa which are associated in some epididymal segments, in many species, but usually not agglutinated after ejaculation (3);

(2) protection of gametes by adsorption of enzyme inhibitors (13); (3) modification of permeability and metabolism, perhaps through variations of cyclic AMP (7).

It is remarkable that all observed modifications (acrosome and cell coat) are achieved in the posterior corpus epididymis where the fertilizing ability of spermatozoa is reached.

References

1 *Barker, L.D.S. and Amann, R.P.:* Epididymal physiology. II. Immunofluorescent analyses of epithelial secretion and absorption, and of bovine sperm maturation. J. Reprod. Fertil. *26:* 319–332 (1971).

2 *Bedford, J.M.:* Changes in the electrophoretic properties of rabbit spermatozoa during passage through the epididymis. Nature, Lond. *200:* 1178–1180 (1963).

3 *Bedford, J.M.:* Observations on some properties of a potent agglutinin in the semen of a fertile rabbit. J. Reprod. Fertil. *22:* 193–198 (1970).

4 *Bedford, J.M.; Cooper, G.W., and Calvin, H.I.:* Post meiotic changes in the nucleus and membranes of mammalian spermatozoa. Proc. Int. Symp. of Genetics of the Spermatozoon, Edinburgh 1972, pp. 69–89.

5 *Bedford, J.M. and Nicander, L.:* Ultrastructural changes in the acrosome and sperm membranes during maturation of spermatozoa in the testis and epididymis of the rabbit and monkey. J. Anat., *108:* 527–543 (1971).

6 *Brown, C.R. and Hartree, E.F.:* Distribution of a trypsin-like proteinase in the ram spermatozoon. J. Reprod. Fertil. *36:* 195–198 (1974).

7 *Casillas, E.R. and Hoskins, D.D.:* Adenyl cyclase activity and cyclic 3', 5' AMP content of ejaculated monkey spermatozoa. Arch. Biochem. *147:* 148–155 (1971).

8 *Fawcett, D.W. and Phillips, D.M.:* Observations on the release of spermatozoa and on changes in the head during passage through the epididymis. J. Reprod. Fertil. Suppl. *6:* 405–418 (1969).

9 *Fléchon, J.-E.:* Modifications ultrastructurales et cytochimiques des spermatozoïdes de lapin au cours du transit épididymaire. Coll. sur Transport, Survie et Pouvoir fécondant des Spermatozoïdes chez les Vertébrés, vol. 26, pp. 115–140 (INSERM, Paris 1973).

10 *Fléchon, J.-E.:* Freeze-fracturing of rabbit spermatozoa, J. Microscopie *19:* 59–64 (1974).

11 *Fléchon, J.-E.:* Validity of phosphotungstic acid staining of polysaccharide (glycogen) at very low pH on thin sections of glycolmethacrylate-embedded material. J. Microscopie (in press).

12 *Fléchon, J.-E. et Morstin, J.:* Localisation des glycoprotéines et des charges négatives et positives dans le revêtement de surface des spermatozoïdes éjaculés de lapin et de taureau. Coll. int. Histochimie (Tours, in press).

13 *Menezo, Y. et Fléchon, J.E.:* Utilisation d'une microtechnique pour la détermination des activités enzymatiques des spermatozoïdes et du plasma séminal chez le lapin et le taureau. C.R. Acad. Sci. *277:* 1037–1040 (1973).

14 *Nicander, L.:* On the regional histology and cytochemistry of the ductus epididymis in rabbits. Acta. morphol. neerl. scand. *1:* 99–118 (1957).

15 *Nicander, L. and Bane, A.:* Fine structure of the sperm head in some mammals, with particular reference to the acrosome and the subacrosomal substance. Z. Zellforsch. mikrosk. Anat. *72:* 496–515 (1966).

16 *Nicolson, G.L. and Yanagimachi, R.:* Terminal saccharides on sperm plasma membranes; identification by specific agglutinins. Science *177:* 276–278 (1972).
17 *Orgebin-Crist, M.-C.:* Studies on the function of the epididymis. Biol. Reprod. Suppl. *1:* 155–175 (1969).
18 *Waites, G.M.H. and Setchell, B.P.:* Physiology of the testis, epididymis and scrotum. Adv. Reprod. Physiol. *4:* 1–64 (1969).
19 *Weil, A.J.:* Antigens of the seminal plasma. J. Reprod. Fertil. Suppl. *2:* 25–34 (1967).
20 *Yanagimachi, R.; Noda, Y.D.; Fujimoto, M., and Nicolson, G.L.:* The distribution of negative surface charges on mammalian spermatozoa. Amer. J. Anat. *135:* 497–520 (1972).
21 *Zaneveld, L.J.D.:* Sperm acrosomal enzymes and their potential use for contraceptive development. Coll. sur Transport, Survie et Pouvoir fécondant des Spermatozoïdes chez les Vertébrés, vol. 26, pp. 435–450 (INSERM, Paris 1973).

Author's address: *J.E. Fléchon,* Institut National de la Recherche Agronomique, Station Centrale de Physiologie animale, *F–78350 Jouy-en Josas* (France)

Sperm Transport in Female Tract of Primates

The Biology of Spermatozoa
INSERM Int. Symp., Nouzilly 1973, pp. 48–55 (Karger, Basel 1975)

Sperm Transport in the Female Tract of Nonhuman Primates

E.S.E. Hafez and S. Jaszczak

Departments of Gynecology-Obstetrics and Physiology, and C.S. Mott Center for Human Growth and Development, Wayne State University School of Medicine, Detroit, Mich.

The physiological, biochemical and immunological aspects of transport of mammalian spermatozoa in the female reproductive tract have been reported (1–9). Discrepancies exist regarding the mechanisms of sperm transport from the site of ejaculation to the site of fertilization in primates. These discrepancies are probably due to anatomical and physiological differences in species studied, and to the variable methods used to study the distribution, survival and fertilizing life of spermatozoa. Several factors are known to influence the ascent of spermatozoa from the vagina to the oviduct, e.g. (a) intrinsic sperm motility; (b) cyclical variations in the biophysical and biochemical characteristics of luminal secretions of the female reproductive tract; (c) contraction of the female reproductive tract as influenced by copulatory stimuli, oxytocin, prostaglandins and other spasmogens in the semen (10–14). The relative importance of these factors is unknown. The penetration of spermatozoa in the cervical mucus is an important factor in sperm ascent.

The purpose of this experiment was to study the factors which affect the distribution of spermatozoa within the female reproductive tract on nonhuman primates.

Materials and Methods

Experimental Animals

Adult females of three species of nonhuman primates were used: crab-eating macaques *(Macaca fascicularis),* rhesus monkey *(Macaca mulatta)* and marmoset monkey (15). Upon arrival, the animals were thoroughly examined, weighed, held in quarantine for 1–2 months, tuberculin-tested and given other medications as needed. The animals were individually housed in cages in ventilated, air-filtered rooms at a constant temperature (about 76 °F), constant humidity (40 %) and daily photoperiod of 12 h of light. Commercial food pellets and water were available *ad libitum*, and were supplemented daily by apples or bananas.

The onset of menstruation was considered as day 1 of the cycle. The quantity and properties (spinnbarkeit, ferning and cellularity) of cervical mucus were recorded daily around the time of ovulation. Samples of cervical mucus were aspirated into a tuberculin syringe. Spinnbarkeit was measured by stretching a thread of mucus between a glass slide and a cover slip. A crystallization test was performed by spreading a thin thread of clear cervical mucus on slide, and the dried mucus was examined under the microscope. The degree of secondary and tertiary branching of the crystals was recorded. Shortly after ovulation, the quantity, spinnbarkeit and ferning of cervical mucus decreased abruptly. Laparoscopy was performed to ensure ovulation.

Blood samples were collected every second day between day 1 and 8 of the cycle; then daily from day 9 until the day that the cervical mucus became scanty without spinnbarkeit and ferning; and, finally, every second day to the beginning of the next cycle.

Breeding and Autopsy

Females with regular menstrual cycles, satisfactory quantity and properties of cervical mucus, and proven ovulatory cycles were bred at the estimated time of ovulation to an experienced male with known semen quality. The female was separated from the male after two ejaculations. Vaginal content was examined for the presence and motility of spermatozoa. The femals were autopsied 1, 3, 6, 16, and 24 h after copulation. Before autopsy the animals were anesthesized with an intravenous injection of 2 ml of ketamine. *In vivo* fixation of the reproductive tract was performed by rapid injection of 5 ml of diabutal, followed by the injection of 30 ml of 10 % neutral formalin into the abdominal aorta. The reproductive tract was dissected out and immersed in 10 % neutral formalin.

In the preliminary experiments of this investigation the survival time of spermatozoa was estimated by postcoital examination of spermatozoa recovered from the cervical canal 1, 12, 24, 48, and 72 h after copulation.

Histology and Scanning Electron Microscopy

Segments of the cervix, uterus and oviduct were processed and viewed by scanning electron microscopy using the technique described by *Hafez* (16). Sagittal and transverse sections at 10 μ from the cervix, uterus and oviducts, were stained with modified Mallory's triple stain.

Counts of Spermatozoa and Leukocytes

A microscopic counting square with a surface area of 0.144 mm^2 was used to count spermatozoa and leukocytes at 70–150 sites in histological sections, taken at random in each segment. Every second or third histological section was used. The same counting square and magnification (× 480) were used for all counts throughout the study.

Hormone Assay

Serum luteinizing hormone (LH) was determined by radioimmunoassay using ^{131}I for iodination procedure. For LH standards both monkeys LH (LERM 907 D) and ovine (LER 1056 C2) were used. Plasma progesterone was determined by the competitive protein binding.

Results

Functional Anatomy of Female Reproductive Tract

The gross anatomical characteristics of the uterus and oviduct are similar in man and macaque species. However, there are quantitative and qualitative differences in the length, shape and diameter of the total body of the cervix, and length shape and diameter of the endocervical canal. In the rhesus monkey and crab-eating macaque, the cervix is relatively large and the cervical canal is obstructed by one main ventral colliculus and a pair of smaller and variably developed dorsal colliculi. Interdigitation of these colliculi result in a peculiar dorso-ventral sinusity of the cervical canal. The cervical mucosa contains finger-like projections, which in certain areas (especially close to external os) have secondary and tertiary branches. These structures of cervical mucosa result in a large surface area of mucus secretory epithelium, with abundant amounts of cervical mucus.

The cervix, uterus and oviducts are lined with different types of nonciliated secretory cells and ciliated cells. In general, secretory cells have a dome-shaped surface covered with numerous microvilli and their cytoplasm contains numerous secretory granules. The cell membrane of secretory cells ruptures with the release of secretory granules into the lumen. During the storage phase, the cell membrane is complete over its luminal surface with its microvilli.

The percentage of ciliated cells in the epithelium varies in different parts of the reproductive tract, being maximal in the fimbriae and oviduct, much lower in the cervix and minimal in the uterus. The ciliated cells are covered with kinocilia which beat rhythmically toward the vagina. The smooth endometrium is characterized by few mucosal folds and numerous openings of the endometrial glands.

Semen Characteristics

The whole semen coagulated upon ejaculation. There were remarkable variations among species of macaque and individuals of the same species in the rate of coagulation and liquefaction of semen *in vivo* and *in vitro*. The weight of ejaculates from the rhesus monkey ranges from 3 to 5 g and upon liquefaction, 0.5–0.7 ml was separated within 20–40 min at 37 °C.

Sperm Migration through the Cervix

Histological studies on sperm distribution in the cervical canal at various intervals from copulation, as well as in the different stages of the menstrual cycle revealed remarkable differences in the accumulation of spermatozoa in different segments. The number of spermatozoa decreased gradually from the external os of the cervix to the uterus. In animals bred at midcycle, maximal gradients of spermatozoa between external and internal ora of the cervix were observed at 1, 3, 6, and 12 h after copulation. By 16 or 24 h after insemination the number of spermatozoa in the cervix significantly declined, and concentration gradients between various portions of the cervix was less evident.

Near ovulation, the migration of sperm in the cervical mucus toward the uterus was mostly at random. Spermatozoa were distributed more or less uniformly in the midcervix and especially in the endocervix. However, spermatozoa were often densely crowded in some strands of mucus passing through the longitudinal axis of the uterus. Some strands of mucus containing large numbers of spermatozoa were directed toward the lumina of cervical crypts or to the uterine cavity.

The extent and depth of sperm penetration was directly related to the quantity and properties of cervical mucus. During the early follicular and luteal phase of the cycle, the cervical canal was occluded by a highly viscid, thick, gelatinous mucus plug, which inhibited the ascent of spermatozoa. Large numbers of spermatozoa, oriented irregularly, were found in the exocervical canal but very few in the midcervix and endocervix and uterus.

Spermatozoa were often found in the cervical crypts. Their number in the cryptal lumina varied with intervals from copulation and with stages of the menstrual cycle. The amounts of spermatozoa within the cervical crypts were positively correlated with the numbers of spermatozoa found in the cervical canal. The number of leukocytes in cervical crypts was usually much lower than the number of spermatozoa.

Sperm Migration in the Uterus

Spermatozoa were sometimes aggregated along certain strands of endometrial mucus in the uterine cavity. Sperm entered the uterine cavity and endometrial glands in two different ways: (a) through the lumen of the cervical canal, and (b) through the longitudinal cervical crypts connecting with some endometrial glands. The latter pathway of spermatozoa offers some advantages, since the uterine lumen contains larger number of leukocytes than the endometrial glands.

The majority of spermatozoa found in the uterus were densely aggregated within the endometrial glands, and only a small number of sperm were present in the uterine cavity. Most of the endometrial glands, however, did not contain any spermatozoa. Reconstruction of glandular patterns of the endometrium from serial histologic sections indicated that some 'routes' of sperm followed the longitudinal direction of the gland whereas others were aggregated in the distal end of the endometrial glands (table I).

Sperm Migration to the Oviduct

Sperm counts per surface area showed much higher values in the vicinity of the uterotubal junction compared with the oviducts (table II). Spermatozoa were predominantly accumulated with the lumina of some longitudinal endometrial glands. This finding may suggest that in nonhuman primates, as in some other species, the uterotubal junction may serve as a barrier for the ascent of sperm. Very few spermatozoa were located in the oviduct due to the continuous migration of spermatozoa from the ostium of oviduct to the peritoneal cavity.

Table I. Number of spermatozoa within the uterus of monkeys

Number of animals	Stage of cycle	Hours p.c. autopsied	Uterine cavity, mean and range	Lumina of endometrial glands, mean and range	Utero-tubal junction, mean and range
1	ovulation	3	8.7 (0–17)	17.8 (3–45)	4.1 (1–23)
1	ovulation	6	17.2 (2–39)	28.3 (2–59)	5.4 (0–21)
3	ovulation	12	9.3 (0–22)	36.2 (5–56)	21.2 (3–49)
1	ovulation	16	3.4 (0–9)	11.5 (0–41)	2.9 (0–11)
1	ovulation	24	2.1 (0–11)	13.1 (2–39)	1.8 (0–13)
1	early follicular	6	0.8 (0–3)	0.1 (0–2)	0
1	midluteal	6	0.1 (0–1)	0	0

Figures represent total number of spermatozoa present in one longitudinal histological section (10 μm) of the uterus.

Table II. Number of spermatozoa within the lumen of oviducts in monkeys bred at the time of ovulation 12, 16, and 24 h after copulation

Hours p.c. autopsied	Number of animals	Nonligated oviducts			Ligated oviducts
		isthmus	ampulla	total	
12	3	37	69	106	271
16	1	21	26	47	not done
24	1	34	18	52	not done

Similarities and Dissimilarities between Human and Nonhuman Primates

Certain morphological and anatomical similarities and dissimilarities were noted in the cervix uteri of women and different species of macaques. The structure of cervix in the rhesus and crab-eating macaque is very different from that of women. The cervix of the patas monkey *(Erythrocebus patas patas)* and baboon *(Papio anubis),* both African primates, is much closer anatomically to women than the rhesus (17–19).

The quantitative and qualitative changes in the concentration of gonadotropins and ovarian steroids in peripheral blood throughout the menstrual cycle of the macaque resemble those of the human cycle. Also the secretion of cervical mucus in the monkey is regulated by ovarian hormones and follows a regular cyclic pattern.

Discussion

Sperm transport in nonhuman primates is influenced by structural characteristics of the cervical and endometrial mucosa; cyclic changes in quantity and quality of cervical mucus; patterns of copulatory behavior; and the rate of coagulation and liquefaction of semen. Semen coagulation, a reaction between the secretions of cranial prostate and the vesicular gland (20, 21), seems to be similar to that in the guinea pig and rat. The enzymes removed from the rat coagulating gland have the ability to coagulate monkey semen. Variations in the liquefaction of the coagulum either by natural or experimental means have attributed to wide variations in reported constituents of the semen and in the sperm concentration of various species of nonhuman primates (22).

The remarkable difference in the population of sperm encountered in various portions of the female monkey's reproductive tract may have different physiologic implications at various stages of the cycle. Shortly before or at the time of ovulation, the presence of massive numbers of spermatozoa in the cervix may serve as a sperm reservoir. It is also possible that massive numbers of spermatozoa are destroyed in the cervical crypts to avoid the migration of excessive numbers of spermatozoa to the uterine lumen. The function of the cervix as a sperm reservoir has been suggested for ruminants and rabbits (13, 23, 24). However, the concept that the cervix functions as a sperm reservoir is not supported by experimental results. It is suggested that a large amount of sperm, present in cervical mucus, consists mainly of inferior quality and morphologically abnormal spermatozoa which are unable to overcome resistance of cervical mucus (25, 26). Further evidence is needed to elucidate the function of the cervix in relation to storage of sperm and the fertilizing potentials of motile sperm stored in the mucus. Remarkable concentration gradients of spermatozoa between lower and upper parts of the cervix observed in the present study may indicate that cervical mucus present within the lower part of the cervix is the first and most important barrier preventing the entry of a vast majority of sperm toward the upper reproductive tract. The cervical mucosa contains significant percentage of ciliated cells primarily near the openings of the cervical crypts. These cells are lined with kinocilia which beat toward the vagina (27). Little is known about the role of kinocilia in the transport of normal and abnormal sperm to the cervical crypts, uterine cavity endometrial glands, and peritoneal cavity.

Dukelow and Chernoff (28) studied the survival of spermatozoa of the rhesus monkey and man in the uteri of various laboratory species. The superiority of the hamster as an experimental animal is clearly shown both for the percent of motile sperm and for the total sperm recovered. The poorest rhesus sperm survival occurred in the mouse, whereas human sperm survival was poorest in the rat. Survival of both species of sperm was low in the guinea pig. It would appear that the spermatozoa of human and nonhuman primates do require capacitation before they are capable of fertilizing an ovum. The time required and direct proof for this capacitation await new developments of research techniques.

References

1 *Davajan, V.; Nakamura, R.M., and Kharma, K.:* Spermatozoan transport in cervical mucus. Obstet. gynec. Surv. *25:* 1 (1970).

2 *Frenkel, D.A.:* Sperm migration and survival in the endometrial cavity. Int. J. Fertil. *6:* 285 (1961).

3 *Hancock, J.L. and McGovern, P.T.:* The transport of sheep and goat spermatozoa in the ewe. J. Reprod. Fertil. *15:* 283 (1968).

4 *Haynes, N.B.:* The influence of the uterine environment on the phagocytosis of spermatozoa; in *Lamming and Amoroso* Reproduction in the female mammal (Butterworth, London 1967).

5 *Howe, R.G.:* Leukocytic response to spermatozoa in ligated segments of the rabbit vagina, uterus and oviduct. J. Reprod. Fertil. *13:* 563 (1967).

6 *Kellerman, A.S. and Weed, J.C.:* Sperm motility and survival in relation to glucose concentration: an *in vitro* study. Fertil. Steril. *21:* 802 (1970).

7 *Menge, A.C.; Tyler, W.J., and Casida, L.E.:* Factors affecting the removal of spermatozoa from the rabbit uterus. J. Reprod. Fertil. *3:* 396 (1962).

8 *Sobrero, A.J. and MacLeod, J.:* The immediate post-coital test. Fertil. Steril. *13:* 184 (1962).

9 *Duijin, C. van, jr.; Voorst, C. van, and Freund, M.:* Movement characteristics of human spermatozoa analysed from kinemicrographs. Europ. J. Obstet. Gynec. *4:* 121 (1971).

10 *Bedford, J.M.:* The saga of mammalian sperm from ejaculation to syngamy; in *Gibian and Plotz* Mammalian reproduction (Springer, Berlin 1970).

11 *Dobrowolski, W. and Hafez, E.S.E.:* Transport and distribution of spermatozoa in the pseudopregnant rabbit. Amer. J. vet. Res. *31:* 2243–2246 (1970).

12 *Banna, A.A. El and Hafez, E.S.E.:* Sperm transport and distribution in rabbit and cattle female tract. Fertil. Steril. *21:* 534–540 (1970).

13 *Hafez, E.S.E.:* Gamete transport; in *Hafez and Evans* Human reproduction. Conception and contraception, chap. 4 (Harper & Row, New York 1973).

14 *Moghissi, K.S.:* The function of the cervix in fertility. Fertil. Steril. *23:* 295–306 (1972).

15 *Hafez, E.S.E.:* Comparative reproduction of nonhuman primates (Thomas, Springfield 1971).

16 *Hafez, E.S.E.:* Scanning electron microscopy of the female reproductive tract. J. Reprod. Med. *9:* 119 (1972).

17 *Banna, A.A. El and Hafez, E.S.E.:* The uterine cervix in mammals. Amer. J. Obstet. Gynec. *112:* 145–164 (1972).

18 *Hafez, E.S.E.:* Comparative anatomy of mammalian cervix; in *Blandau and Moghissi* Biology of the cervix, chap. 2 (University of Chicago Press, Chicago 1973).

19 *Ovadia, J.; McArthur, J.W.; Kopito, L., and Ulfelder, H.:* The cervical mucus secretion of the bonnet monkey *(M. radiata).* Anatomical basis and physiological regulation. Biol. Reprod. *5:* 127–145 (1971).

20 *Hoskins, D.D. and Patterson, D.L.:* Prevention of coagulum formation with recovery of motile spermatozoa from rhesus monkey semen. J. Reprod. Fertil. *13:* 337 (1967).

21 *Wagenen, G. van:* The coagulating function of the cranial lobe of the prostate gland in the monkey. Anat. Rec. *44/66:* 411–421 (1936).
22 *Dukelow, W.R.:* Semen and artificial insemination; in *Hafez* Comparative reproduction of nonhuman primates (Thomas, Springfield 1971).
23 *Gibbons, R.A. and Mattner, P.E.:* The chemical and physical characteristics of the cervical secretion and its role in reproductive physiology; in *Sherman* Pathways to conception (Thomas, Springfield 1971).
24 *Mattner, P.E.:* The cervix and its secretion in relation to fertility in ruminants; in *Blandau and Moghissi* Biology of the cervix (University of Chicago Press, Chicago 1973).
25 *Bergman, P.:* Sperm migration and its relation to the morphology and motility of spermatozoa. Int. J. Fertil. *1:* 45–54 (1955).
26 *Botella-Llusia, J.:* Methods for determining the type and degree of spermatic motility; in *Westin and Wiqvist* Proc. 5th Wld Congr. Fert. Steril., Stockholm 1966. Int. Congr. Ser. No. 133, pp. 636–644 (Excerpta Medica, Amsterdam 1967).
27 *Hafez, E.S.E. and Kanagawa, H.:* Ciliated epithelium in the uterine cervix of the macaque and rabbit. J. Reprod. Fertil. *28:* 91–94 (1972).
28 *Dukelow, W.R. and Chernoff, H.N.:* Primate sperm survival and capacitation in a foreign uterine environment. Amer. J. Physiol. *216:* 682–686 (1969).
29 *Elstein, M.; Moghissi, K.S., and Borth, R.* (eds.): Cervical mucus in human reproduction (WHO Publication, Scriptor, Copenhagen 1973).
30 *Jaszczak, S. and Hafez, E.S.:* Sperm migration through the uterine cervix in the macaque during the menstrual cycle. Amer. J. Obstet. Gynec. *115:* 1070–1082 (1973).

Authors' address: Prof. *E.S.E. Hafez* and Dr. *S. Jaszczak*, Department of Gynecology and Obstetrics, C.S. Mott Center for Human Growth and Development, Wayne State University School of Medicine, *Detroit, MI 48201* (USA)

The Biology of Spermatozoa
INSERM Int. Symp., Nouzilly 1973, pp. 56–62 (Karger, Basel 1975)

Studies on Sperm Migration in the Human Female Genital Tract

H.B. Croxatto, A. Faundes, M. Medel, S. Avendaño, H.D. Croxatto, C. Vera, J. Anselmo and L. Pastene

Laboratorio Endocrinología, I.C.B., Universidad Católica de Chile, and Hospital Barros Luco-Trudeau, Unidad Obstetricia-Ginecología, Santiago

Introduction

Little is known about the fate of spermatozoa beyond the internal os of the cervix in women. The literature on this subject up to 1969 was critically reviewed by *Ahlgren* (1), and additional work has been reported subsequently by various groups (2–8).

The bulk of the data indicates that spermatozoa can reach the site of fertilization within the first 2 h after insemination and that they can stay there motile for nearly 48 h. Plastic IUCDs do not seem to offer a major obstacle for sperm migration, while at least partial interference with the process is observed under continuous treatment with progestogens. Phagocytosis by leukocytes apparently helps to dispose of surplus spermatozoa, specially in the endometrial cavity with an IUCD in place (3). The fact that large numbers of spermatozoa pass through the oviducts into the peritoneal cavity could represent another way to eliminate spermatozoa from the genital tract, but it also leaves open for consideration the possibility that fertilization might occur in the peritoneal cavity, as suggested earlier by *Horne and Thibault* (9).

The relative roles of sperm motility, secretions of the genital tract, smooth muscle and ciliary activity, and semen constituents in sperm transport and fertilizing ability, are at present a matter of speculation. Similarly, mechanisms that permit preferential migration of the X or Y chromosome (10) bearing sperm, or preferential migration of normal over abnormal spermatozoa (11), have been proposed and stand as possible sophistications of sperm migration in humans. Last but not least, the systemic and local neuroendocrine regulation of the processes mentioned above also add to the list of questions which constitute our current challenge.

The observations described here were done initially to define the distribution of spermatozoa in the genital tract several hours after intercourse in normal untreated women and in patients under various contraceptive treatments.

Materials and Methods

The patients admitted to these studies were of high parity and had given recent proof of being fertile. They were divided into four groups according to treatment.

Group I: 20 normal women who received no treatment during the cycle in which surgery was performed. The majority of them had been using an IUCD since their last delivery or abortion and the device was removed from 1 to 3 months before the study. During this period they used condom, vaginal douche, suppositories or periodic abstinence.

Group II: Composed of five women treated with megestrol acetate orally, one tablet or one capsule daily, containing either 0.25, 0.5, or 0.75 mg. Details concerning this treatment have been published by *Avendaño et al.* (12).

Group III: Five women who were protected by the sequential pill method. The medication consisted of 16 tablets of mestranol 0.1 mg followed by 5 tablets of mestranol 0.075 mg plus esthynodiol diacetate 1 mg. They were operated on the day they took the sixteenth estrogen pill.

Group IV: Composed of seven women wearing an IUCD with copper (13). All women were instructed not to have intercourse until the night prior to admission for surgery and their compliance with this advice was checked at frequent intervals by microscopic examination of cervical mucus. The surgical procedure consisted of laparotomy, aspiration of fluid from the pouch of Douglas, bilateral salpingectomy, ovarian biopsy, flushing of the endometrial cavity with 5 cm^3 of saline, and endometrial biopsy. The endometrial cavity was flushed by transparietal puncture of the fundus after salpingectomy. Each oviduct was flushed with 5 ml of Tyrode solution through a needle inserted in the isthmus.

The fluids were examined under a dissecting microscope to search for eggs and to remove tissue fragments. In a few instances, the flushings from both oviducts were combined but usually they were treated separately. Fluids heavily contaminated with blood were not used.

To search for spermatozoa, the fluids were centrifuged at 2,000 r.p.m. for 15 min in conical tubes. The sediments obtained were treated with eosin, distributed in slides, dried, mounted and screened entirely at 625–800 times magnification by phase contrast microscopy.

In order to establish the rate of recovery of spermatozoa by this method, four oviducts obtained from three patients that did not have intercourse during the cycle of surgery were used. 10 μl of a diluted fresh or frozen semen sample were injected into the lumen of the isthmus *in vitro* and, after 20 min, the procedure described above was followed.

The recovery was fairly constant and independent of the initial number of spermatozoa. About 70 % were lost throughout the procedure but no effort was done to find out at which step the loss occurred. The number of spermatozoa reported in the tables are the actual numbers found, instead of corrected figures. The endometrial and corpus luteum biopsies were dated according to established criteria (14, 15). A more detailed account on methods was published elsewhere (16).

Results

The three segments – uterine cavity, oviducts and peritoneal cavity – were examined in only four cases of group I and the search for sperm was negative in three patients of this group.

Tubal eggs were recovered in Op21, Op26, Op36, and Op59. Motile spermatozoa were seen around and in the cumulus in Op21. Nonmotile spermatozoa were seen in the cumulus in Op26. No association with sperm was noted in the other two.

The distribution of spermatozoa in normal women whose uterine cavity and oviducts were examined 8–15 h postcoitum is shown in table I.

The recovery of spermatozoa in the four women whose uterine cavity was not examined is shown in table II.

The relationship between number of sperm recovered and stage of the menstrual cycle is shown in table III.

The findings in patients under contraceptive treatment are shown in table IV. The proportion of cases with spermatozoa present in each segment of the four groups is shown in table V.

Table I. Distribution of spermatozoa in normal women whose uterine cavity and oviducts were examined 8–15 h postcoitum

	Uterine cavity	Oviducts			Peritoneal cavity
		right	left	both	
Uterine cavity only (31 %)					
Op17 (12)	39	0	0	0	–
Op24 (14:30)	24	–	–	0	0
Op34 (12)	7	0	0	0	–
Op67 (11)	2	0	–	0	–
Oviducts only (8 %)					
Op61 (9)	0	–	–	2	–
Uterine cavity and oviducts (61 %)					
Op26 (37 and 8)	1,380	75	122	197	148
Op37 (12:30)	1,490	162	114	276	844
Op47 (15)	1,379	287	325	612	–
Op51 (11:30)	151	98	92	190	73
Op55 (10:50)	1,573	2,184	–	?	–
Op59 (12)	2,905	–	10	?	–
Op65 (11)	115	–	–	16	–
Op68 (12:30)	24	8	59	67	–

Figures in parentheses after each case indicate hours postcoitum.

Table II. Number of spermatozoa recovered in normal women whose uterine cavity was not examined

	Oviducts			Peritoneal cavity
	right	left	both	
Op21 (10:30)	4,980	–	?	–
Op22 (11)	3	5	8	–
Op28 (144 and 17)	–	–	2	0
Op48 (12:50)	0	417	417	9

Figures in parentheses after each case indicate hours postcoitum.

Table III. Number and distribution of spermatozoa in relation to the stage of the menstrual cycle in normal women

	Uterine cavity	Mean number per oviduct
Follicular phase		
Op17	39	0
Op37	1,490	138
Op47	1,397	306
Op55	1,573	1,092[1]
Op65	115	8
Luteal phase		
Op32 Day 1	0	0
Op51 Day 1	151	95
Op59 Day 1	2,905	10
Op68 Day 2	24	33
Op48 Day 3	0	208[1]
Op61 Day 3	0	1
Op34 Day 5	7	0
Op24 Day 6	24	0
Op67 Day 9	2	0

1 Only one oviduct examined.

Discussion

The technique employed to recover and count spermatozoa from the oviducts gave a low yield representing from one-fifth to one-third of the actual number.

Table IV. Distribution of spermatozoa in women under contraceptive treatment

	Uterine cavity	Oviducts			Peritoneal cavity
		right	left	both	
Megestrol acetate					
Op14 (24 and 6:30)	515	–	–	0	–
Op15 (168 and 10:30)	0	–	–	0	–
Op29 (6)	0	–	0	?	–
Op30 (144:120 and 6:30)	80	10	18	28	–
Op43 (12)	0	0	0	0	0
Mestranol					
OpS1	0	31	0	31	–
OpS2	0	0	8	8	–
OpS3	5	–	17	?	–
OpS4	–	109	–	?	–
OpS6	–	12	65	77	–
IUCD with copper					
Op31 (13:30)	0	0	0	0	–
OpCu1 (11)	10	0	0	0	–
OpCu7 (13)	4	–	–	–	–
OpCu8 (14)	–	0	3	3	–
OpCu13 (19:30)	76	2	2	4	2
OpCu17 (12:30)	43	2	3	5	–
OpCu18 (12)	1	–	4	?	0

Figures in parentheses after each case indicate hours postcoitum.

Table V. Proportion of cases with spermatozoa present in the different segments studied

Group	Endometrial cavity	Oviducts	Peritoneal cavity
Control	12/16[1]	13/20	4/6
Megestrol acetate	2/5	1/5	0/1
Mestranol	1/3	5/5	0/0
T-Cu	5/6	4/6	1/2

1 Number of cases with spermatozoa present number of cases examined.

The number of recovered in the control group showed a remarkably wide range, from 0 to 5,000. From 1,000 to 5,000 spermatozoa were found in 38 % of the endometrial cavities and in 10 % of the oviducts examined in this group. Such large numbers found beyond the internal os of the cervix had been reported previously only in women with laterally closed oviducts (1), a condition that is definitely excluded here. It must be kept in mind though that these spermatozoa were not necessarily at the site of fertilization.

A strong correlation between the number of spermatozoa recovered and the stage of the menstrual cycle was apparent. During the follicular phase the median for the endometrial cavity was 1,397 spermatozoa and for each oviduct 138, whereas during the luteal phase, the corresponding values were 7 and 1, respectively.

In treated women the recovery was generally low as compared with the controls and this was more pronounced at the level of the oviduct in the group treated with megestrol acetate and the one with the T-Cu. The physiologic significance of a smaller number of spermatozoa and the contribution of this effect to the contraceptive mechanism is uncertain. What should be stressed is that none of these treatments completely abolished sperm migration to the Fallopian tubes, which is in contrast with the report of *Zañartu et al.* (4) and in agreement with that of *Gutierrez-Najar et al.* (17).

It is possible that the effect of megestrol acetate reported here might be due entirely to interference with sperm migration through the cervical mucus (18, 19). The effect of the T-Cu, in turn, could be ascribed to the known toxic effect of copper ions on human spermatozoa (20, 21).

Acknowledgements

This work was supported in part by The Population Council grants to H.B.C., A.F., M.M. and S.A. The assistance of Mrs. *M. Delgado* in the preparation of tissue sections and that of Miss *C. Llados* and Mrs. *M. Ortiz* in screening the slides, is greatly appreciated.

References

1 *Ahlgren, M.:* Migration of spermatozoa to the Fallopian tubes and the abdominal cavity in women, including some immunological aspects; med. Diss., Lund (Studentliterature, Lund 1969).

2 *Moyer, D.L.; Rimdusit, S., and Mishell, D.R.:* Sperm distribution and degradation in the human female reproductive tract. Obstet. Gynec. *35:* 831–840 (1970).

3 *Sagiroglu, N.:* Phagocytosis of spermatozoa in the uterine cavity of woman using intrauterine device. Int. J. Fertil. *16:* 1–14 (1971).

4 *Zañartu, J.; Pupkin, M.; Rosemberg, D.; Guerrero, R.; Rodriguez-Bravo, R.; Garcia-Huidobro, M., and Puga, J.A.:* Effect of oral continuous progestogen therapy in microdosage on human ovary and sperm transport. Brit. med. J. *ii:* 266–269 (1968).

5 *Faundes, A.; Croxatto, H.; Medel, M.; Vera, C.; Avendaño, S., and Anselmo, J.:* Sperm migration in the female genital tract. 7th Wld Congr. on Fertility and Sterility, Tokyo 1971.

6 *Ahlgren, M.:* Number of spermatozoa in the ampulla of the human Fallopian tube. 7th Wld Congr. on Fertility and Sterility, Tokyo 1971.

7 *Kesserü, E.; Larrañaga, A.; Hurtado, H., and Benavides, G.:* Fertility control by continuous administration of d-Norgestrel 0.03 mg. Int. J. Fertil. *17:* 17–27 (1972).

8 *Settlage, D.S.F.; Motoshima, M., and Tredway, D.:* Sperm transport from the external cervival os to the Fallopian tubes in women. A time and quantitative study. 29th Ann. Meet. American Fertility Society, San Francisco 1973.

9 *Horne, H.W. and Thibault, J.P.:* Sperm migration through the human female reproductive tract. Fertil. Steril. *13:* 135–139 (1962).

10 *Shettles, L.B.:* Factors influencing sex ratios. Int. J. Gynec. Obstet. *8:* 843–647 (1970).

11 *Bergman, P.:* Sperm migration and its relation to the morphology and motility of spermatozoa. Int. J. Fertil. *1:* 45–54 (1955).

12 *Avendaño, S.; Tatum, H.J.; Rudel, H.W., and Avendaño, O.:* A clinical study with continuous low doses of megestrol acetate for fertility control. Amer. J. Obstet. Gynec. *106:* 122–127 (1970).

13 *Medel, M.; Guiloff, E.; Toscanini, C.; Pastene, L.; Rivera, M.; Zipper, J., and Gómez-Roger, C.:* Clinical evaluation of an IUD with 200 mm² of copper. A new approach in human contraception; in *Goldsmith and Snowden* Family Planning Research Conference, Multidisciplinary Approach, Exeter 1971 (Exerpta Medica, Amsterdam 1972).

14 *Noyes, R.W.; Hertig, A.T., and Rock, J.:* Dating the endometrial biopsy. Fertil. Steril. *1:* 3–25 (1950).

15 *Corner, G.W.:* The histological dating of the human corpus luteum of menstruation. Amer. J. Anat. *98:* 377–401 (1956).

16 *Croxatto, H.B.; Faundes, A.; Medel, M.; Avendaño, S.; Croxatto, H.D.; Vera, C.; Anselmo, J., and Pastene, L.:* Studies on sperm migration in the human female genital tract; in Transport, survie et pouvoir fécondant des spermatozoïdes chez les vertébrés (INSERM, Paris 1973).

17 *Gutierrez-Najar, A.; Giner-Velasques, J., and Martinez-Manautou, J.:* Role of cervical mucus in contraception with the continuous chlormadinone acetate method; in *Sobrero and Lewitt* Advances in planned parenthood, vol. 4, pp. 97–102 (Excerpta Medica, Amsterdam 1969).

18 *Mason, B.A.; Cox, H.J.E.; Mason, D.W., and Grant, V.:* Clinical and experimental studies with low doses of megestrol acetate. Postgrad. med. J. Suppl. Dec. pp. 45–48 (1967).

19 *Lebech, P.E.; Svendsen, P.A., and Ostergaard, E.:* The effect of small doses of megestrol acetate in the cervical mucus. Int. J. Fertil. *15:* 65–76 (1970).

20 *White, I.C.:* The toxicity of heavy metals to mammalian spermatozoa. Austr. J. exp. Biol. *33:* 359–366 (1955).

21 *Loewit, K.:* Immobilization of human spermatozoa with iron. Basis for a new contraceptive? Contraception *3:* 219–224 (1971).

Author's address: Prof. *H. Croxatto,* Universidad Católica de Chile, Instituto de Ciencias Biológicas, Laboratorio de Endocrinología, Casilla 114-D, *Santiago* (Chili)

The Biology of Spermatozoa
INSERM Int. Symp., Nouzilly 1973, pp. 63–73 (Karger, Basel 1975)

Sperm Transport and Survival in Women with Special Reference to the Fallopian Tube

M. Ahlgren, Kerstin Boström and R. Malmqvist

Department of Obstetrics and Gynecology, Lund, University Hospital and Department of Forensic Medicine, University of Lund, Lund

Introduction

Available relevant data regarding sperm physiology in the human Fallopian tube is limited. After artificial insemination or coitus, spermatozoa rapidly reach the oviduct at time of ovulation: within 5 min following insemination, sperm are present at the fimbriated end (33), and proceed into the abdominal cavity (1). Survival time and duration of fertilizing capacity is unknown. The number of spermatozoa in the ampulla of the oviduct is low – estimated at not more than 200 (2). It is also not known whether there is a qualitative morphological selection of spermatozoa on their way through the female genital tract. In the present study, the morphology and survival time of spermatozoa in the human Fallopian tube will be reported.

Review of Literature

There are few pertinent reports on sperm migration to the Fallopian tubes and abdominal cavity in women. Table I presents a complete list of all reports that have appeared in the literature. A more detailed review is recently published by *Ahlgren et al.* (3).

Materials and Methods

Women seeking advice for infertility in the years 1970–2 were included in the present study. The group of women called infertile also contains an unknown number of potentially fertile women where the infertility is due to the husband. A couple was considered infertile when pregnancy had not occurred during at least one year of normal sexual endeavor. All infertile couples were scheduled for routine gynecological examination of the woman, basal

Table I. Survey of studies on sperm migration to the oviducts and abdominal cavity in women

Authors	Number and classification of cases	Methods	Results: oviducts, abdominal cavity (ac)
Birch-Hirschfeld (1892)	prostitute cadaver	postmortem examination	motile sperm 14–16 h after coitus
Dührssen´ (1893)	1 chronic salpingitis	bilateral salpingectomy	motile sperm 3.5 weeks after coitus
Nürnberger (1920)	3 cases of ovarian cysts	unilateral salpingectomy	nonmotile sperm in two cases, 2 motile sperm in one case 6, 13 and 14–15 days after coitus
	1 pyosalpinx (unilat)	bilateral salpingectomy	1 motile and 1 nonmotile sperm in open oviduct 13 days after coitus
Hoehne (1921)	7 laparotomy cases	salpingectomy 1–7 days after coitus	1 nonmotile sperm at 24 h p.c. at infundibulum
Belonoschkin (1939)	5 sterilizations	tubal smears	no sperm 2.5–6.5 days after coitus
Brown (1944)	6 excised uteri and oviducts	perfusion *in vitro* with saline (3 uteri); wrapping in saline compresses (3 uteri)	sperm migration through the whole genital tract (1 uterus perfused, 2 not perfused) in 68–71 min after application of cervical semen pool
Davis (1947)	?	?	sperm in oviduct 3 h after coitus
Farris (1950)	7 patients requiring hysterectomy, 2 weeks coitus abstinence prior to operation	hysterectomy + salpingectomy,	motile sperm in oviducts up to 43 h after insemination
Rubenstein et al. (1951)	49 cases of fibroids, polyps or endometriosis, unknown fertility, no restriction from coitus before insemination	hysterectomy + salpingectomy, flush with saline after operation	sperm found in 25 cases 30 min to 50 h after insemination

Table I (continued)

Authors	Number and classification of cases	Methods	Results: oviducts, abdominal cavity (ac)
Horne and Audet (1958)	13 infertile women	laparotomy with cul-de-sac aspiration	motile sperm in oviduct 60 h after coitus, active sperm in 2/13 cases 36 h after coitus (ac)
Horne and Thibault (1958)	15 infertile women	(a) laparotomy with tubal washing from the cervix	sperm in 10/15 tubal washings sperm in 6/15 cases 24 h after coitus (ac)
	6 infertile women	(b) transvaginal cul-de-sac aspiration	motile sperm 3/6 cases, inactive sperm in 1/6 cases 24 h after coitus (ac)
Kelly (1962)	25 infertile women, 72 h preoperative restriction from coitus	transvaginal cul-de-sac aspiration	sperm in 2/25 cases 1–8 h after coitus (ac)
Malkani and Sujan (1964)	4 fertile women with Margulies intrauterine spiral	salpingectomy: tubal washing in two parts	sperm in 4/4 patients on the day after coitus
Sjövall (1964)	24 infertile women	laparoscopy	oviductal sperm in 1 case, peritoneal (ac) sperm in 2 cases, 3.5–10 h after coitus
Morgenstern et al. (1966)	13 vaginal hysterectomies, 10 with IUD, 3 controls	extirpation and flushing of one tube at operation	1–187 sperm in 5 cases, 1–2 days after coitus
Zanartu (1966)	12 fertile women on steroid therapy for fertility control	oviductal flush at laparotomy about 20 h after coitus in cases undergoing sterilization or other surgery	sperm in 5/5 on cyclic estrogens or sequential steroids; no sperm in 7 women on combined or progestagen therapy
	4 fertile women with IUD		sperm in 4/4 women

Table I (continued)

Authors	Number and classification of cases	Methods	Results: oviducts, abdominal cavity (ac)
Moghissi (1968)	4 excised uteri with tubes from fertile women with preinvasive cancer or uterine prolapse	uterus and tubes enveloped in saline sponges	no sperm above cervix which was in semen pool up to 18 h
Gutierrez-Najar et al. (1969)	39 young fertile women of which 24 were on continuous chlormadinone	culdoscopy with flushing of the lateral end of the tubes	sperm in 14/15 cases in control group and 2/24 cases in treated group 8–12 h after coitus
Moyer et al. (1970)	15 vaginal hysterectomies for pelvic relaxation	oviductal flush	sperm in 1/15 cases 36 h after coitus
Ahlgren (1969, 1971)	177 infertility cases; 3 fertile women; preoperative restriction from coitus min. 4 days	laparoscopy and laparotomy; oviduct and cul-de-sac aspiration	sperm in 53/177 infertile and 1/3 fertile women; number of sperm less than 200 in open ampullae – in sactosalpinges up to 23,000; sperm in 22/42 infertile and 1/3 fertile women (ac) migration time 1 h 50 min (ac)
Faundes et al. (1971)	27 sterilizations	flush of removed tubes	20 % sperm recovery 7–37 h after coitus
Settlage et al. (1973)	8 sterilizations; preoperative restriction from coitus 11–14 days	removed tubal segments rinsed in serum	sperm present at fimbriated end 5 min following artificial insemination

body temperature recording to indicate the presence and day of ovulation, hysterosalpingography to demonstrate tubal patency, endometrial biopsy in selected cases and spermiogram of the husband. If pregnancy occurred during this part of the investigation the case was not included in the present material. Those cases where the husband was found to be sterile or seriously subfertile were also excluded. After the above-listed program was fulfilled the woman was scheduled for postcoital laparoscopy or laparotomy. Laparoscopy was performed when the hysterosalpingogram was normal or nonconclusive to find possible endometriosis or small adhesions around ovaries and oviducts which might disturb the mechanism by which eggs are picked up into the infundibulum (7). Such pathological findings could be expected in 40–50 % of the cases (1, 31). Laparotomy was performed when the hysterosalpingogram indicated bilateral sactosalpinges without patency to the abdominal cavity. The couple was instructed to have coitus at the presumed ovulation time and also at a given time before admission to the hospital. Coitus was restricted for at least 4 days before the 'admissional' intercourse. The reasons for these 4 days restriction were twofold. Sperm survival in the human oviduct has been reported to be limited to 50–60 h (17, 32). Thus with at least 4 days (i e. about 96 h) of abstinence from coitus before the investigational coitus, it should be fairly certain that no spermatozoa from the preceding coitus would be found in the oviducts. Moreover, 4 days seems to be a common spontaneous interval of coitus.

Two groups of patients were selected. In one group of 19 infertile *patients chosen for studies of the morphology of oviductal spermatozoa* all but one had sactosalpinges. The one with laterally open oviducts was originally not scheduled for sperm morphology studies at 14 h after coitus but because she was found to have an exceptionally high number of spermatozoa in the oviductal secretion she was also included in the group. In another group of 20 infertile patients where the purpose of the investigation was to find out the *survival time of the spermatozoa in the oviduct* the interval between 'admissional' coitus and examination at operation was between 60 and 113 h. The study was designed to begin at 60 h after coitus for the reasons given above. In this group all but five patients had open, normal oviducts. Five had sactosalpinges and one of these was also included in the 'morphological' group at 60 h postcoitum. Two patients had earlier surgical removal of one oviduct.

After adequate anesthesia was initiated, a Sims-Huhner test was performed as described by *Ahlgren* (1). As the number and motility of spermatozoa in the female secretions is often different in different microscopic fields it was considered more meaningful to judge the whole sample from each slide than to give numbers and motility per high power field. The grading of the number was thus simplified to either none, few or several. The motility pattern was classified as either nonmotile, motile without progression, or motile with progression.

At laparoscopy all the fluid (10–20 ml) in the Douglas' pouch was aspirated, the Fallopian tubes were inspected and grasped by a special organ forceps (27) at the fimbriated end and catheterized by means of a plastic tube – Bardic® premature infant feeding tube, size 5, length 15 in (fig. 1–2). The external diameter of the plastic tube (1–1.5 mm) prevented it from entering the isthmus of the oviduct. The feeding tube was slowly withdrawn from the ampulla of the oviduct under simultaneous gentle suction by a syringe which was attached to the end of the plastic tube. The plastic tube was clamped on both sides of the tubal mucus specimen obtained and kept so until analyzed for spermatozoa in the laboratory after the end of the operation.

In the laparotomy cases with sactosalpinges the ampulla was punctured with an intravenous cannula and the fluid of the ampulla aspirated into a plastic syringe of adequate size.

The contents of all feeding tubes were exsufflated on a glass slide as soon as possible, covered with a slip (18 × 18 mm), and immediately scanned for spermatozoa under the phase contrast microscope (× 100). The amount of oviductal fluid obtained from one side was usually the proper volume for one cover slip.

The Douglas' fluids were centrifuged at about 3,000 r.p.m. for 10 min, the supernatant withdrawn, and the sediment examined for the presence of spermatozoa.

The fluids obtained from the oviducts closed at the fimbriated end were analyzed in a similar way. When the presence of sperm was noted in a small sample of the fluid, the rest was centrifuged at about 3,000 r.p.m. for 10 min. The sediment was used to produce a smear on a glass slide, which was fixed in 95 % of ethanol for 30 min and stained with

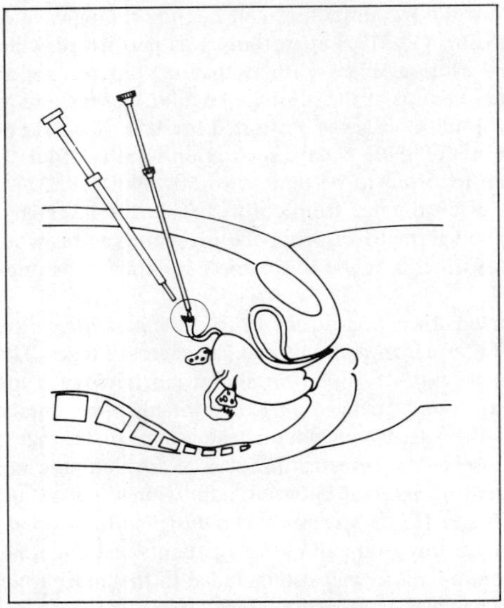

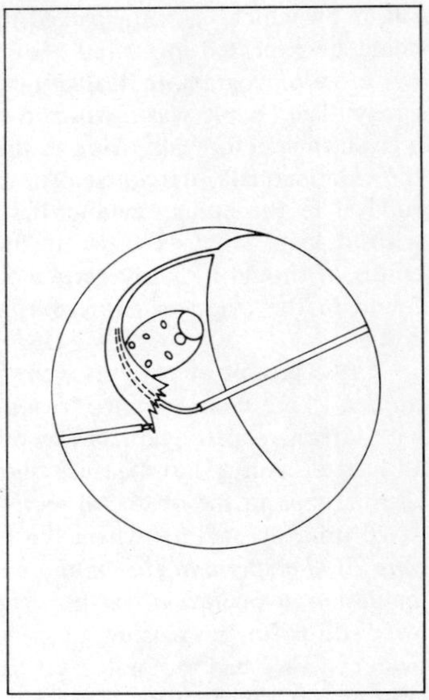

1, 2

Fig. 1. Survey of laparoscopic procedure.

Fig. 2. Encircled detail of figure 1 showing catheterization of the ampulla of the oviduct.

Table II. Distribution of spermatozoa with normal morphology in the human ejaculate, uterus and oviduct

Case No.	Hours p.c.	Ejaculate (E)		Uterus (U)		Oviduct (O)		Difference	
		%	n	%	n	%	n	U-E	O-E
1	35	85	300	–	–	89	100	–	4
2	16	75	300	–	–	79	246	–	4
3	11	86	300	95	600	95	250	9*	9
4	18	85	300	98	500	98	400	13**	13**
5	14	87	300	–	–	91	124 (Dx)	–	4
						97	500 (Sin)	–	10*
6	15	87	300	–	–	97	500	–	10*
7	10	86	100	–	–	97	300	–	11
8	60	83	300	96	300	96	30	13*	13

* Difference probably significant ($p < 0.05$). ** Difference significant ($p < 0.01$).
n = Number of spermatozoa judged, i.e. in the ejaculates 300 (100 in case No. 7) consecutively observed spermatozoa from the seminal smears; in the uterine and oviductal smears all spermatozoa were judged.

hematoxylin-eosin. For comparison an ejaculate was obtained from the husband within 1 week after the operation of the woman. The morphology of the spermatozoa was examined according to *Boström and Andersson* (8). The judgement of the appearance of spermatozoa was performed by Dr. *Boström.*

Results

Morphology of Spermatozoa

The distribution of spermatozoa with normal morphology from the oviducts, uterus and ejaculate is shown in table II. This table includes only the eight cases, in which the number of spermatozoa was large enough for an adequate judgement of the appearance of the spermatozoa. In seven of these cases the spermatozoa were recovered from sactosalpinges. In case number 5 the finding of several hundred spermatozoa in normal oviducts was an exception from the common finding of a maximum of 200 spermatozoa. In the other 11 cases of the group with sactosalpinges only few or no spermatozoa were found.

Only three uterine specimens were suitable for morphological judgement. The rest contained too much blood. Uterine spermatozoa differed from ejaculated spermatozoa by a higher percentage of normal morphology. The difference was statistically significant. The percentages of spermatozoa with normal mor-

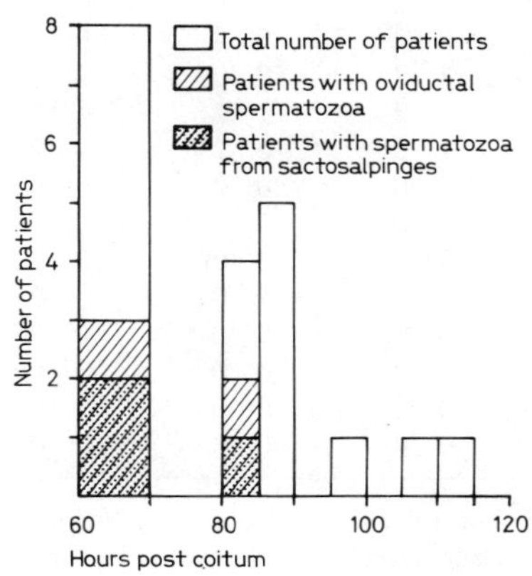

Fig. 3. Survival time of spermatozoa in the human oviduct (ampulla).

phology from the oviducts were consistently higher than in the ejaculates although this difference was statistically significant only in three cases.

Survival Time of Spermatozoa

The recovery of spermatozoa from open oviducts and a few sactosalpinges is shown in figure 3 and table III. Motile progressive spermatozoa were recovered from the ampulla of the Fallopian tube up to 85 h after coitus. No spermatozoa

Table III. Recovery of spermatozoa from the human uterus, oviduct and abdominal cavity at different times after coitus

Case No.	Hours p.c.	Recovery of spermatozoa								Remarks
		cervix		uterine cavity		ampulla		abdominal cavity		
		n	motil.	n	motil.	n	motil.	n	motil.	
1	60	1	2	1	2	0	0	0	0	
2	60	2	2	2	2	2	2	1	2	open sactosalpinx
3	62	0	0	0	0	0	0	0	0	further investigation[1]
4	62	0	0	0	0	0	0	0	0	ovulation[2]
5	63	1	2	2	2	1	2	0	0	open sactosalpinx
6	63	0	0	2	2	2	2	1	2	
						1	1			
						1	0			
7	64	0	0	0	0	0	0	0	0	ovulation[2]
8	66	1	2	0	0	0	0	0	0	
9	83	0	0	0	0	0	0	0	0	
10	84	0	0	2	2	0	0	0	0	sactosalpinx
11	85	1	0	1	2	1	2	1	2	
12	85	2	2	1	2	2	2	0	0	sactosalpinx
13	86	0	0	0	0	0	0	0	0	
14	87	0	0	0	0	0	0	0	0	sactosalpinx
15	87	0	0	0	0	0	0	0	0	
16	88	1	0	0	0	0	0	0	0	
17	89	0	0	0	0	0	0	0	0	
18	99	0	0	0	0	0	0	0	0	
19	107	0	0	0	0	0	0	0	0	
20	113	0	0	0	0	0	0	0	0	

Number of sperm: 0 = none; 1 = few; 2 = several sperm in sample. Sperm motility: 0 = immotile; 1 = motile without progression; 2 = motile with forward progression.

1 Later, further investigations revealed poor postcoital tests a few hours after coitus. Husband was eventually found to have a high titre of sperm agglutinins.

2 Ovulation had probably preceded coitus by at least 48 h.

were recovered from oviducts or uterine cavities 86–113 h from coitus, i.e. eight patients. Spermatozoa were recovered from the cervical mucus of only one of the eight women tested, and those spermatozoa were nonmotile.

Discussion

The observation that spermatozoa with normal morphology are more common in the oviduct than in the ejaculate is new. *Bergman* (5) demonstrated a lower number of abnormal forms of spermatozoa in the uterus than in the ejaculate, which fits in with the result of *Odeblad* (29) who found that spermatozoa must be normally shaped in order to progress in between the micelles of the cervical mucus. The results of the present study seem to indicate that spermatozoa are selected in the cervix in such a way that morphologically normal spermatozoa reach the site of fertilization to a greater extent than abnormal ones. This could be explained by the fact that morphologically abnormal spermatozoa often have a decreased motility (22). On the other hand, the results of this study show that some morphologically abnormal spermatozoa also reach the oviduct, although to a lesser extent. Whether this is related to the occurrence of spontaneous abortions is not known. It has been stated that a high concentration of abnormal spermatozoa in the ejaculate could be connected with a high rate of spontaneous abortion (19). In the discussion of a selection in the cervical mucus between spermatozoa with normal and abnormal appearance the entire explanation might lie in the fact that in a seminal smear from the ejaculate both motile and nonmotile spermatozoa are observed but in the smear from the uterus and the oviducts almost all spermatozoa are motile. The mechanisms for the selection thus should be that the active sperm transport is not working for those spermatozoa from the ejaculate which are nonmotile at the time when they reach the external os of the cervical canal.

Horne and Audet (17) reported motile spermatozoa from the Fallopian tube of one patient 60 h after intercourse. In the present study motile spermatozoa have been found in the ampulla of the normal oviduct up to 85 h after coitus. In all patients with a longer observation time, up to 113 h after coitus, no spermatozoa were found. This does not preclude the possiblity of a longer survival time. Active spermatozoa have been found within the cervical mucus 7 days after insemination with donor semen (30). In the present study only a few immotile spermatozoa were found in the cervical mucus of one patient out of eight examined between 85 and 113 h after coitus.

Although motile spermatozoa within the ampulla of the oviduct have been observed 85 h after coitus it is not certain whether these spermatozoa are fertile at this time. *Knaus* (21) demonstrated that fertility ceased before motility in the rabbit, and in the human no direct observation is reported, but from experience with artificial insemination with donor sperm, *Farris* (13) has concluded the fertile life of sperm to be 24 h. To analyze this problem one would need to test the fertilizing ability of the spermatozoa in an *in vitro* system of the type established by *Edwards* (12) and co-workers.

References

1 *Ahlgren, M.:* Migration of spermatozoa to the Fallopian tubes and the abdominal cavity in women, including some immunological aspects; med. diss., Lund (Studentlitteratur, Lund 1969).

2 *Ahlgren, M.:* Number of spermatozoa in the ampulla of the human Fallopian tube; in Proc. 7th Wld Congr. on Fertility and Sterility, Tokyo and Kyoto 1971, Int. Congr. Ser. No. 278 (Excerpta Medica, Amsterdam 1973).

3 *Ahlgren, M.; Boström, K., and Malmqvist, R.:* Sperm transport and survival in women with special reference to the Fallopian tube; in *Thibault and Hafez* Transport, Survie et pouvoir fécondant des spermatozoïdes chez les vertébrés (INSERM, Paris 1973).

4 *Belonoschkin, B.:* Das Verhalten der menschlichen Spermatozoen im weiblichen Organismus. Münch. med. Wschr. *86:* 847–850 (1939).

5 *Bergman, P.:* Sperm migration and its relation to the morphology and motility of spermatozoa. Int. J. Fertil. *1:* 45–54 (1955).

6 *Birch-Hirschfeld* (1892); cited in *Hoehne and Boehne:* Über die Lebensdauer der homologen und heterologen Spermatozoen im weiblichen Genitalapparat und in der Bauchhöhle. Zbl. Gynäk. *38:* 5–20 (1914).

7 *Blandau, R.J.:* Gamete transport – comparative aspects; in *Hafez and Blandau* The mammalian oviduct, pp. 129–162 (University of Chicago Press, Chicago 1969).

8 *Boström, K. and Andersson, L.:* Creatinine phosphokinase relative to acid phosphatase, lactate dehydrogenase, zinc and fructose in human semen with special reference to chronic prostatitis. Scand. J. Urol. Nephrol. *5:* 123–132 (1971).

9 *Brown, R.L.:* Rate of transport of spermia in human uterus and tubes. Amer. J. Obstet. Gynec. *47:* 407–411 (1944).

10 *Davis, M.E.:* Personal commun.; in *Farris* Human fertility and problems of the male, p. 20 (Author's Press, White Plains 1950).

11 *Dührssen, Z.:* Geburtsh. Gynäk. *27:* 216 (1893).

12 *Edwards, R.G.:* Physiological aspects of human ovulation, fertilization and cleavage. J. Reprod. Fertil. Suppl. *18:* 87–101 (1973).

13 *Farris, E.J.:* Human fertility and problems of the male (Author's Press, White Plains 1950).

14 *Faundes, A.; Croxatto, H.; Medel, M.; Vera, C.; Avendano, S., and Anselmo, J.:* Sperm migration in the human female genital tract. Proc. 7th Wld Congr. on Fertility and Sterility, Tokyo and Kyoto 1971. Int. Congr. Ser. No. 278 (Excerpta Medica, Amsterdam 1973).

15 *Gutierrez-Najar, A.; Giner-Velazques, J., and Martinez-Manautou, J.;* Role of cervical mucus in contraception with the continuous chlormadinone acetate method; in *Sobrero and Lewit* Advances in planned parenthood, vol. 4, pp. 97–102 (Excerpta Medica, Amsterdam 1969).

16 *Hoehne, O.:* Zur Physiologie der Befruchtung beim Menschen. Zbl. Gynäk. *45:* 1047–1048 (1921).

17 *Horne, H.W., jr. and Audet, C.:* 'Spider cells', a new inhabitant of peritoneal fluid. Obstet. Gynec. *11:* 421–423 (1958).

18 *Horne, H.W., jr. and Thibault, J.P.:* Sperm migration through the human female reproductive tract. Fertil. Steril *13:* 135–139 (1962).
18a *Horne, H.W., jr. and Thibault, J.P.:* Sperm migration through the human female reproductive tract. Fertil. Steril. *13:* 444–447 (1962).
19 *Jones, G.E.S. and Delfs, E.:* Endocrine patterns in term pregnancies following abortion. J. amer. med. Ass. *146:* 1212–1218 (1951).
20 *Kelly, J.V.:* Myometrial participation in human sperm transport. A dilemma. Fertil. Steril. *13:* 84–92 (1962).
21 *Knaus, H.:* Zur Physiologie der Spermatozoen. Arch. Gynäk. *151:* 302–329 (1932).
22 *MacLeod, J.:* Semen quality in relation to infertility and to contraception; in *Sobrero and Lewit* Advances in planned parenthood, vol. 4, pp. 145–149 (Excerpta Medica, Amsterdam 1969).
23 *Malkani, P.K. and Sujan, S.:* Sperm migration in the female reproductive tract in the presence of intrauterine devices: a preliminary report. Amer. J. Obstet. Gynec. *88:* 963–964 (1964).
24 *Moghissi, K.S.:* Human and bovine sperm migration. Fertil. Steril. *19:* 118–122 (1968).
25 *Morgenstern, L.L.; Orgebin-Crist, M.-C.; Clewe, T.H.; Bonney, W.A., and Noyes, R.W.:* Observations on spermatozoa in the human uterus and oviducts in the chronic presence of intrauterine devices. Amer. J. Obstet. Gynec. *96:* 114–118 (1966).
26 *Moyer, D.L.; Rimdusit, S., and Mishell, D.:* Sperm distribution and degradation in the human female reproductive tract. Obstet. Gynec. *35:* 831–840 (1970).
27 *Nordvall, S.:* An instrument to aid laparoscopic diagnosis. Acta obstet. gynec. scand. *49:* 111–112 (1970).
28 *Nürnberger, L.:* Klinische und experimentelle Untersuchungen über die Lebensdauer der menschlichen Spermatozoen. Mschr. Geburtsh. *53:* 87–101 (1920).
29 *Odeblad, E.:* The functional structure of human cervical mucus. Acta obstet. gynec. scand. *47:* suppl. 1, pp. 57–79 (1968).
30 *Perloff, W.H. and Steinberger, E.: In vivo* survival of spermatozoa in cervical mucus. Amer. J. Obstet. Gynec. *88:* 439–442 (1964).
31 *Peterson, E.P. and Behrman, S.J.:* Laparoscopy of the infertile patient. Obstet. Gynec. *36:* 363–367 (1970).
32 *Rubenstein, B.B.; Strauss, H.; Lazarus, M.L., and Hankin, H.:* Sperm survival in women. Motile sperm in the fundus and the tubes of surgical cases. Fertil. Steril. *2:* 15–19 (1951).
33 *Settlage, D.S.F.; Motoshima, M., and Tredway, D.R.:* Sperm transport from the external cervical os to the Fallopian tubes in women: a time and quantitation study. Fertil. Steril. *24:* 655–661 (1973).
34 *Sjövall, A.:* The possibility of an extended postcoital test by laparoscopy in cases of sterility. Proc. 1st Int. Symp. on Gynec. Celioscopy, Palermo 1964.
35 *Zanartu, J.:* Effect of natural and synthetic sex steroids in cervical mucus penetration and ascent of spermatozoa; in *Westin and Wiqvist* Proc. 5th Wld Congr. on Fertility and Sterility, Stockholm 1966. Int. Congr. Ser. No. 133, pp. 704–709 (Excerpta Medica, Amsterdam 1967).

Authors' address: M. *Ahlgren,* Assoc. Prof., Dr. *Kerstin Boström* and Dr. *R. Malmqvist,* Department of Obstetrics and Gynecology, University Hospital, *S–22185 Lund* (Sweden)

The Biology of Spermatozoa
INSERM Int. Symp., Nouzilly 1973, pp. 74–82 (Karger, Basel 1975)

Sperm Transport from the Vagina to the Fallopian Tubes in Women

Diane S.F. Settlage, Masanobu Motoshima and Donald Tredway
Department of Obstetrics and Gynecology, Los Angeles County and University of Southern California Medical Center, Los Angeles, Calif.

Introduction

Studies of sperm transport in the human female are few. It probably does not occur except during the late proliferative and ovulatory portions of the menstrual cycle (13, 18). Data on subprimate species give roles in sperm transport to sperm motility, uterine and oviductal contractility, coital thursting with oxytocin release, seminal plasma constituents and possibly oviductal cilliary movements (18, 19). Animal data also suggest great depletion in the numbers of deposited sperm which actually reach the oviduct, and variation in transport times which may range from 5 min to several hours depending on the species and the author.

Clinical and *in vitro* studies in the human again support the importance of sperm motility in passage through the cervix (3, 5, 10, 14, 17, 18). However, passage of inert particles has been described from the vagina to the oviduct (2, 6). Similarly, uterine and cervical contractions producing negative pressure, and 'insuck' theory, are suggested by some clinical studies though this is denied by others (11, 18).

The possibility also exists that seminal plasma contains a factor or factors increasing cervical mucus solubility (15, 16, 18) or enhancing any role the uterus might play in passage of sperm through the cervical mucus (18, 19). Certainly prostaglandin E_2, present in large quantities in seminal plasma, increases uterine and possibly oviductal contractility (18, 19). Whether or not this is essential, enhancing, or inhibiting to sperm transport through the uterus to the oviducts is unknown, as is the level of the female reproductive tract where an obvious effect might be exerted.

Despite the bits of information described, data on the physiological characteristics of sperm transport in the human female reproductive tract are scant under any conditions. *Rubenstein et al.* (15) demonstrated motile sperm in the cervix, uterine cavity and oviducts by 30 min after insemination. No correlation was made to ovulatory status of patients, quality of inseminates, or phase of the cycle when the study was done. All patients had uterine or tubal pathology. *Brown* (4) demonstrated sperm transport in excised pathologic utero-oviductal preparations exposed to ejaculates by 69 min, confirming *in vitro* capacity for transport. Again the variables mentioned in *Rubenstein*'s work were uncontrolled. *Moghissi* (2), however, was unable to demonstrate sperm transport in a similar study utilizing normal excised organs, suggesting that proposed intrinsic factors may be extremely important. *Horne and Thibault* (8) studied presence or absence of sperm in peritoneal fluid collected at laparoscopy 24 h after coitus as a prognostic tool for cases of infertility. *Ahlgren* (1), more recently, found lower numbers of sperm in the oviductal ampullae of normal women, than in those with clinical and laparoscopic evidence of fimbrial oviductal obstruction at undefined times following coitus. Sperm numbers found in tubal washings or aspirations of women with hydrosalpinges were much higher and may indicate that sperm pass through the tube and into the peritoneal cavity in normal women. *Faundes et al.* (7) utilized oviductal, endometrial and peritoneal washings in patients for sterilization and found sperm in 80 % of patients in some part of these washings 7–37 h after coitus. No attempt to systematically study sperm transport in the human or nonhuman primate is reported under normal or altered conditions which might clarify any potential mechanisms. The present study concerns itself with characterizing human sperm transport under optimal, controlled, assessable conditions. It is complete only for the early features of sperm transport at this writing.

Methods

Subjects selected for this study were women who specifically requested surgical sterilization and volunteered for the project. All subjects were regularly menstruating women, proven fertile, with no history and/or evidence of reproductive organ abnormalities. Additionally, no subjects received hormonal medications or utilized an intrauterine device for a minimum of one menstrual cycle preceding their participation in the study. Any otherwise eligible patient for whom the study requirements constituted sub-optimal management for social or medical reasons peculiar to her, or expressed preference for another surgical approach was excluded from the sample.

As the survival time for sperm in the human Fallopian tube is not exactly known, all subjects were restricted from coitus for from 11 to 14 days preoperatively. Daily examinations of cervical mucus and vaginal secretions during this time confirmed the absence of sperm. Previous reports confirm the detection of sperm in cervical mucus up to 96 h after coitus, and in vaginal secretions for up to 36 h (2, 10, 18). Since practical evaluation of sperm transport must be accomplished during the time when conception is likely, the midcycle status of subjects was determined by the clinical elastoviscosity features of cervical mucus (ferning and spinnbarkeit) and by serum estradiol determinations begun on day 10 of

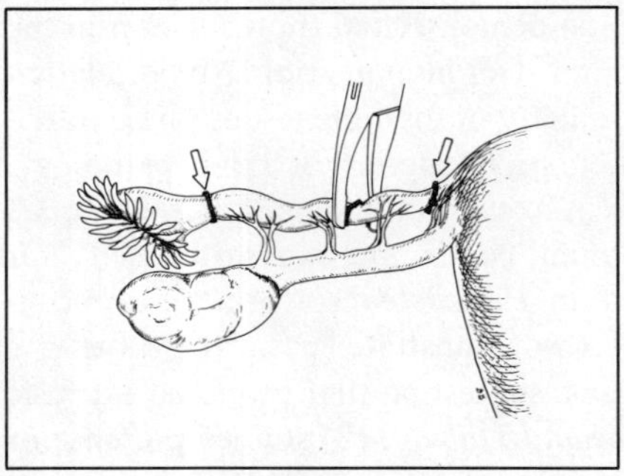

1

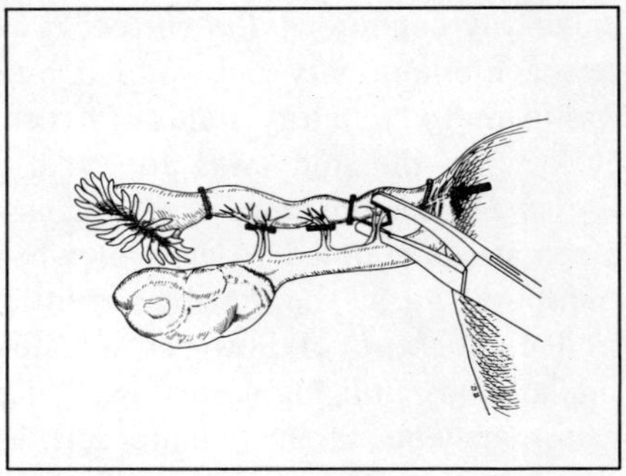

2

Fig. 1. Regional clamping of tubes.
Fig. 2. Technique of salpingectomy.

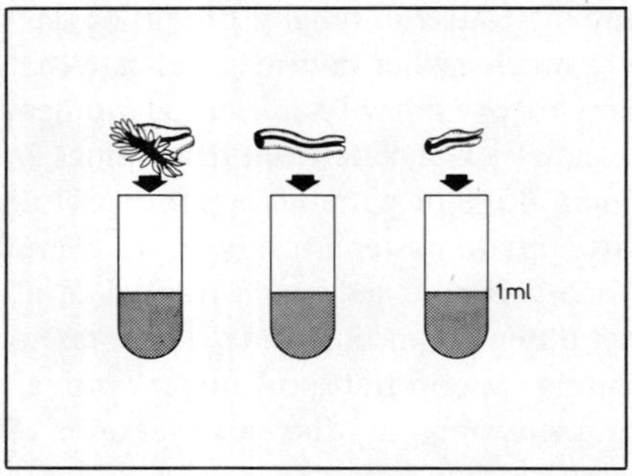

3

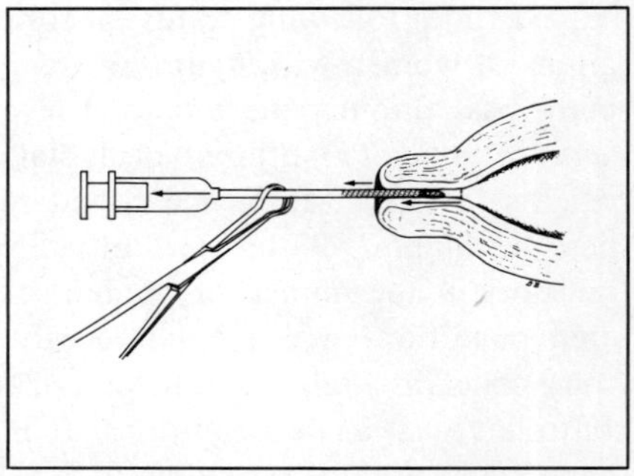

4

Fig. 3. Tubes divided regionally then segments opened and washed.
Fig. 4. Aspiration of cervical mucus prior to and as catheter advanced to the internal cervical os.

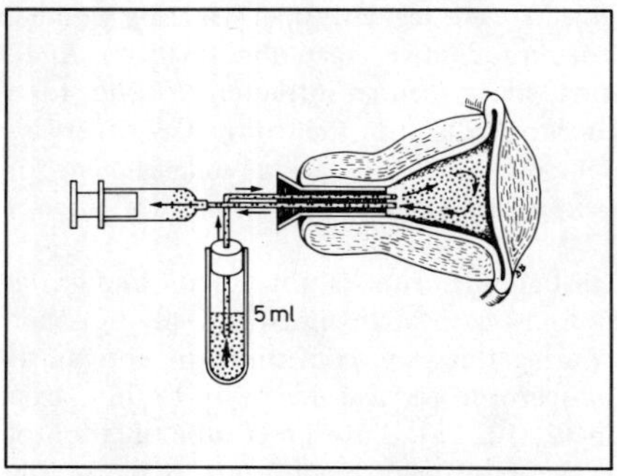

5

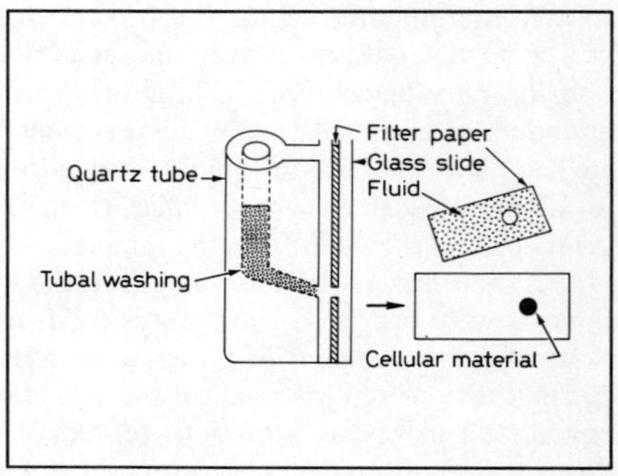

6

Fig. 5. Endometrial cavity wash.
Fig. 6. Cytocentrifugation.

the operative menstrual cycle. The study was done within 36 h of the estradiol preovulatory peak for each subject.

Surgical excision of both Fallopian tubes was accomplished at predetermined times following insemination with fresh, whole, high quality ejaculates which was done by placing a syringe near the external cervical os and gently expelling the inseminate in a pulsatile fashion to mimic coitus. Cervical insemination was not done. The seminal pool remained in the proximal vagina throughout the procedure.

The surgical approach utilized a small horizontal abdominal incision at the pubic hair line, or a prior lower abdominal incision. Following visualization of the uterus and tubes, a small Kelly clamp was placed at the tubo-uterine junction, the isthmic-ampullary junction and the ampullo-fimbrial junction of each tube. Salpingectomy was accomplished by placing multiple Kelly clamps beneath the junction of the tube and the broad ligament, initially severing the tube from the fimbriae ovaricae and then from the uterus at the serosal surface of the uterus. The remaining broad ligament was ligated around the separate clamp pedicles without interference with the utero-ovarian vascular plexus. The time of clamping was different for each tube in each subject to minimize differences between women in our results. In subjects studied before 45 min, insemination was done under a consistent balanced anesthesia. In subjects studied at later times, insemination was done before anesthesia, and both tubes were clamped at the same time. Two subjects, yielding four Fallopian tubes, have been or will be studied at the same time interval.

Following excision, the Fallopian tubes were divided into fimbrial, ampullary, and isthmic segments. These were opened and then vigorously rinsed in test tubes containing 1 ml of serum or normal saline. The tubal segments were removed and fixed for microscopic examination to confirm their histological normalcy. The technique is illustrated in figures 1–3.

The cervical mucus column was removed after the salpingectomy procedure by the technique of *Davajan and Kunitake* (5) without contacting vaginal contents. In this fashion, cervical mucus present at the tip of the tubing represents that at the level of the internal cervical os, and that closer to the syringe represents mucus at the external os. An endometrial cavity washing was then performed utilizing Gravlee jet washes of 5 ml of N saline in 2.5 ml aliquots, followed by a dilation and curettage (fig. 4, 5). Scrupulous care was taken throughout these procedures to avoid tubal segment contamination with sperm and to minimize contamination in the cervical mucus and endometrial specimens.

Donors are healthy presumed fertile male volunteers with known high quality ejaculates who have not ejaculated for 72 h prior to the study. Ejaculates were obtained by masturbation and collected directly into a sterile bottle maintained at body temperature. The total ejaculate volume was measured and inseminated by 30–45 min after collection. A 0.5-ml aliquot was immediately assessed for quality of sperm motility, density per ml, sperm morphology, and the total number of sperm inseminated and the total number of motile sperm inseminated was calculated.

Rinses of the tubal segments, the endometrial and curettage washings were kept carefully isolated from each other and evaluated for sperm content. Rinses were divided into 0.5-ml portions and two drops of 2 % saponin added to lyse red blood cells. The portions were then cytocentrifuged at 500 rpm for 50 min (fig. 6). During this procedure, cellular material was deposited on glass slides in a 0.5-cm circular area and the fluid was absorbed into filter paper. The slides were stained with Giemsa according to the May-Grumwald technique (11). After drying, slides were cover slipped and scanned under a phase contrast magnification (× 400). Any sperm identified was confirmed by examination at × 1,000 oil immersion magnification. No sperm were counted that could not be positively identified.

The cervical mucus column was assessed for the presence of motile sperm at the level of the internal os by direct observation in the polyethylene tubing at × 400 magnification. It was then dissolved in either saturated fructose solution or DTE and the concentration of sperm in the column was determined by standard sperm counting chamber technique (16).

Using these methods, studies of known numbers of sperm added to segments of excised tubes show recovery of sperm from the tubal rinse before cytocentrifugation as 60 % by counting chamber assessment. Recovery of tubal sperm after cytocentrifugation and staining is 2 %, with good reproducibility (coefficient of variation is 50 %). Significance of difference in results is calculated by the Student's *t* test. Numbers and ratios reported are observed, and are not corrected for recovery. Endometrial washing variation on recovery studies is very high and must be interpreted with caution.

Results

All subjects met the study requirements of proven fertility, no gynecological pathology, sperm negative for at least 8 days prior to surgery, and were midcycle by our criteria. Inseminate quality was high. The range of sperm numbers was from 59 to 900 × 10^6 total sperm, 60 to 75 % and 3 to 4 + motility, and from 89 to 97 % normal forms. No sperm were found in tubes removed before insemination, or in tubes removed 2 min after insemination. Sperm were present in the oviducts by 5 min following insemination in the two subjects studied at that interval. Additionally, sperm were distributed throughout all segments of the oviducts at all time intervals at which sperm were found. The total numbers of oviductal sperm found related directly to the numbers of sperm in the inseminate between 15 and 45 min after insemination, but not at later times (fig. 7). Determination of the ratio of sperm in the oviduct to the sperm inseminated and to the motile sperm inseminated revealed a significant difference between the numbers found at 5 min from those at 10 min ($p < 0.05$), and from those found at 10 min to those found between 15 and 45 min ($p < 0.01$).

From 15 to 45 min after insemination, the ratios remained constant indicating no further increase in sperm number with increasing length of time from insemination. Indeed, sperm numbers declined dramatically by 6 h following insemination. None were found in the one subject studied at 48 h. At no other time interval beyond 2 min postinsemination, were sperm absent from the oviducts. Calculation of the percent of sperm found in each segment for each studied oviduct, and then combination of the percentages for all oviducts studied, revealed that consistently the greatest number of sperm were present in the fimbria.

No sperm were found in the endometrial or curettage washings until 90 min after insemination. The numbers of uterine sperm then increased to 24 h but were absent by 48 h after insemination. Ratios of uterine to inseminated and cervical mucus sperm are presented in figure 8. Fragmentation of sperm and the presence of leukocytes was noted at 12 h after insemination and were markedly increased at 24 h.

7 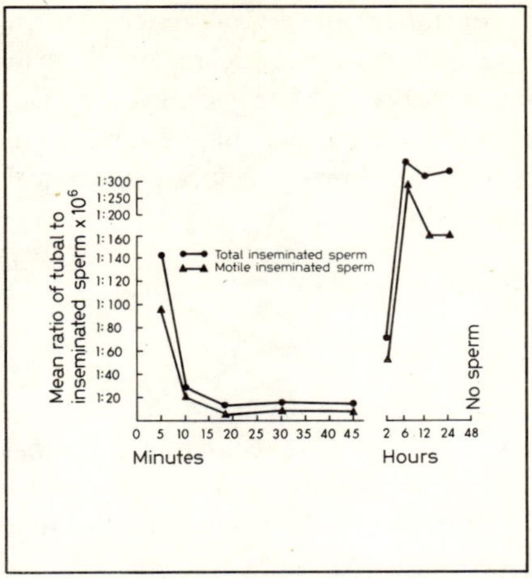

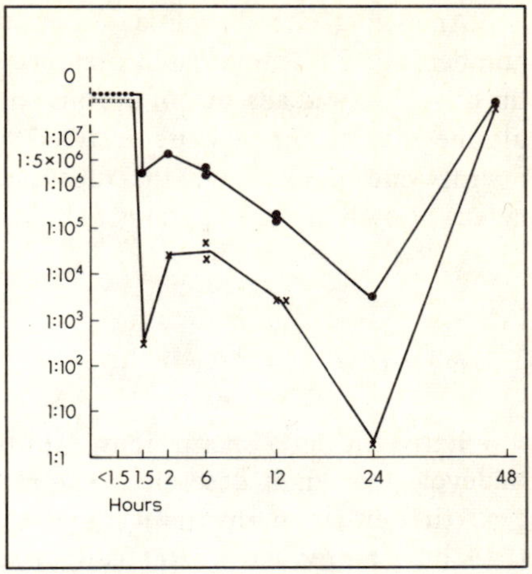 8

Fig. 7. Relationship of tubal sperm to inseminated sperm.
Fig. 8. Sperm ratios: uterine to cervical (x); uterine to inseminate (•).

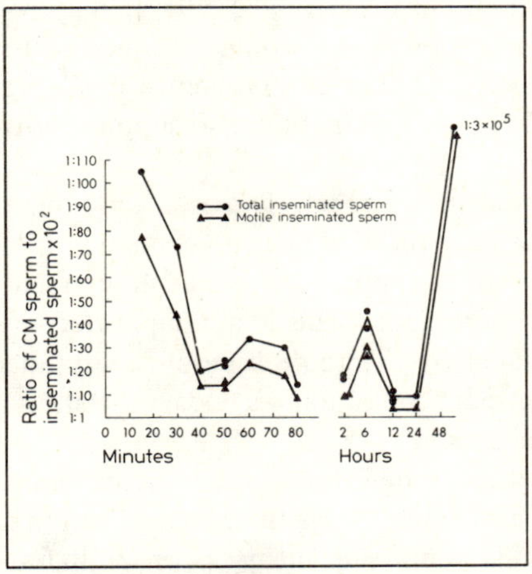

Fig. 9. Relationship to inseminated to cervical mucus sperm.

Sperm found in cervical mucus could be related directly to the sperm inseminated. The ratios of cervical mucus sperm to the motile sperm inseminated seems to be more consistent than to the total sperm inseminated and is fairly constant from 15 min to 24 h after insemination with a mean and standard deviation of $1:15 \neq 9.5 \times 10^2$ (fig. 9).

Analyzing our overall data we note constancy in cervical mucus sperm numbers for 24 h, very rapid presence of tubal sperm with maximal numbers in the first 2 h, and absence of uterine sperm until 1.5–2 h after insemination. The uterine sperm numbers rise as tubal sperm numbers decline for the next 24 h. Sperm were absent from the tubes and uterus and greatly reduced in cervical mucus at 48 h in the one subject studied to this time.

Discussion

Early and later sperm transport utilizing a reproducible technique in normal midcycle women is described. The rapidity of sperm travel to the oviduct suggests that either sperm travel more rapidly *in vivo* than they do *in vitro,* or that the female reproductive tract assists their passage (4, 18).

Recovery of more sperm in the fimbria than in other segments of the oviduct is counter to an assumption that sperm remain in the ampulla till fertilization occurs. It strongly supports, as does the constancy of sperm numbers found between 15 and 45 min after insemination, a theory that sperm traverse the oviduct and are discharged into the peritoneal cavity (12, 19). If they remained in the oviduct, the numbers of recovered sperm would be expected to increase as the time from insemination increased. The observation of declining sperm numbers with increasing time following insemination also supports this premise.

The study subjects differ in one important respect. Those at early time periods were inseminated under anesthesia, and those at later times were not. If the female reproductive tract assisted sperm transport, one would infer from other species that anesthesia would inhibit transport. This is not supported by our data. Conversely, the women inseminated from 2 to 48 h before anesthesia and surgery might also show sperm transport inhibition secondary to stress. These possibilities are currently impossible to assess.

Surprisingly, it appears that the presence of identifiable sperm in the endometrial cavity is not a feature of early sperm transport in the human. *Jaszczak* (9) describes subsurface endometrial glands containing many sperm with few sperm in the cavity of the nonhuman primate. Such might be the case in the human as well. The possibility of poor recovery and hence erroneous interpretation is real and requires additional study in the early periods after insemination. Sperm are recovered at later times from the endometrial cavity, but they increase in number while tubal sperm decline. It would appear that most endometrial cavity sperm do not ascend higher. The fragmentation of sperm and identifiable leukocytes at later time intervals supports theories that sperm induce endometrial leukocytic response.

Total numbers of motile sperm in cervical mucus and the relationship of these to the inseminate and the time from insemination have not been previously described. Beyond 15 min, we see no real increase or decrease in cervical mucus sperm numbers. Several possibilities exist to explain this. The ascendency rate of sperm into the uterus and Fallopian tubes may be relatively slight enough that we cannot yet determine a change. Or, the rate of vaginal sperm penetration into cervical mucus might equal the rate of sperm ascending higher in this early time period. Certainly, the increase of endometrial sperm numbers with increasing time following insemination suggests a reservoir function for cervical mucus.

The preliminary data presented here offer enticing possibilities regarding the phenomenon of sperm transport in the human female reproductive tract. Confirmation of observations, extension of data, improved technique and further characterization requires additional study before definitive mechanisms can be determined.

References

1 *Ahlgren, M.:* Number of spermatozoa in the ampulla of the human Fallopian tube. 7th Wld Congr. on Fertility and Sterility, Tokyo 1971, abstr. 41.

2 *Behrman, S. and Kistner, R.:* Progress in infertility (Little Brown, Boston 1968).

3 *Boer, C.H. de:* Transport of particulate matter through the human female genital tract. J. Reprod. Fertil. *28:* 295 (1972).

4 *Brown, R.L.:* Rate of transport of spermia in human uterus and tubes. Amer. J. Obstet. Gynec. *47:* 407 (1944).

5 *Davajan, V. and Kunitake, G.M.:* Fractional *in vivo* and *in vitro* examination of post-coital cervical mucus in the human. Fertil. Steril. *20:* 197 (1969).

6 *Egli, G.E. and Newton, M.:* The transport of carbon particles in the human female reproductive tract. Fertil. Steril. *12:* 151 (1961).

7 *Faundes, A.; Croxatto, H.; Medel, M.; Vera, C.; Avendans, S., and Anselmo, J.:* Sperm migration in the female genital tract. 7th Wld Congr. on Fertility and Sterility, Tokyo 1971, abstr. 39.

8 *Horne, H.H., jr. and Thibault, J.P.:* Sperm migration through the human female reproductive tract. Fertil. Steril. *13:* 135 (1962).

9 *Jaszczak, S.:* Migration of sperm in the cervix and uterus of non-human primates from cervical mucus in human reproduction. WHO Coll., Geneva 1972, p. 33.

10 *Macleod, J.; Masten, F.; Silberman, C., and Sobrero, A.J.:* The post-coital and post-insemination cervical mucus and semen quality. Steril. Fertil. *10:* 41 (1959).

11 *Mann, T.:* Biochemistry of semen and of the reproductive tract (J. Wiley & Sons, Chichester 1964).

12 *Moghissi, K.S.:* Human and bovine sperm migration. Fertil. Steril. *19:* 119 (1968).

13 *Moyer, D.L.; Rimdusit, S., and Mishell, D.L., jr.:* Sperm distribution and degradation in the human female reproductive tract. Obstet. Gynec. *35:* 831 (1970).

14 *Perloff, W.H. and Steinberger, E.: In vitro* penetration of cervical mucus by spermatozoa. Fertil. Steril. *14:* 231 (1963).

15 *Rubenstein, B.B.; Strauss, H.; Lazarus, M.L., and Hankin, H.:* Sperm survival in woman. Fertil. Steril. *2:* 15 (1951).
16 *Saga, M.; Nakamura, R., and Settlage, D.:* Unpublished.
17 *Schwarz, R.:* Some factors modifying sperm progression. Fertil. Steril. *2:* 15 (1955).
18 *Sobrero, A.J.:* Mechanisms concerned with conception (Pergamon Press, Oxford 1963).
19 *Thibault, C.:* Physiology and physiopathology of the Fallopian tube. Int. J. Fertil. *17:* 1 (1972).

Authors' address: Prof. *Diane S.F. Settlage,* Dr. *M. Motoshima* and Dr. *D. Tredway,* Division of Reproductive Biology, Women's Hospital, Room 2–911, 1400 Mission Road, *Los Angeles, CA 90033* (USA)

The Biology of Spermatozoa
INSERM Int. Symp., Nouzilly 1973, pp. 83–92 (Karger, Basel 1975)

Human Seminal Plasma and Sperm Survival and Transport

A Brief Review

R. Eliasson and C. Lindholmer

Reproductive Physiology Unit, Department of Physiology, Karolinska Institutet, Faculty of Medicine, Stockholm

The relationship between the composition of the human seminal plasma and sperm survival and transport has been given little attention. However, it seems likely that a fluid with such a complex composition as the human seminal plasma should have functions in addition to being a vehicle for the spermatozoa through the final part of the male genital tract and providing them with nutriment.

The first reports that the motility of the human spermatozoa could be dependent on the seminal plasma seem to be by *Hotchkiss* (1) and *Rozin* (2) who noted that in some instances transfer of spermatozoa with poor motility from their own seminal plasma into homologous seminal plasma markedly improved the motility. It has also been noted that semen samples with biochemical signs of prostatic dysfunction more frequently than normal specimens have spermatozoa with impaired motility and viability (3).

To study the effect of fluids from the various accessory genital glands on sperm survival one would prefer epididymal spermatozoa and pure fluids from each of the glands. In human experiments this is for obvious reasons possible only under exceptional conditions. On the other hand, it is well known that during the emission and ejaculation the accessory genital glands discharge their secretions in a specific sequence, i.e. the first portion of ejaculate is mainly prostatic fluid while the later portions mainly have a vesicular origin. The bulk of spermatozoa almost always appears in the first part of ejaculate but spermatozoa can be recovered from all portions of a split ejaculate (4, 5). From a functional point of view these cells can be regarded as 'ampullar' and/or 'epididymal' spermatozoa suspended in prostatic and vesicular fluids, respectively (fig. 1). Acid phosphatases are secreted from the prostate and fructose is a marker for the seminal vesicles in man. The relative concentrations of these – or other organ-specific factors – will therefore in most cases reflect the proportion of

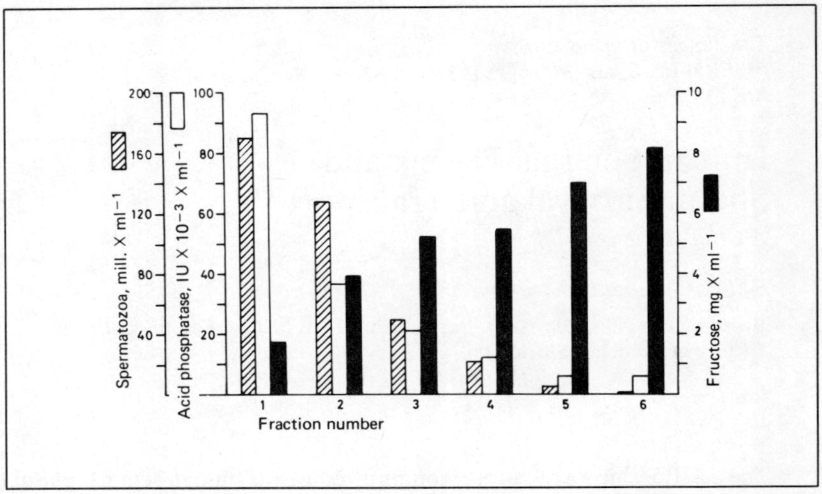

Fig. 1. Distribution of spermatozoa, acid phosphatase activity, and fructose in the various fractions of a split ejaculate. The first fraction has the main part of the prostatic fluid and spermatozoa and the last fraction is almost pure vesicular fluid.

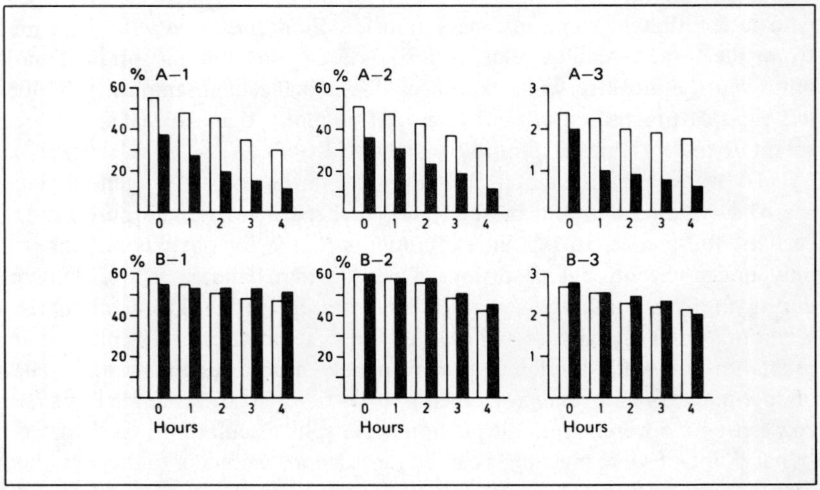

Fig. 2. Sperm motility and survival in the first (open bars) and second half (closed bars) of ejaculates over a 4-hour period. Section A: untreated spermatozoa in their natural milieu; section B: untreated spermatozoa. Albumin was added to the second half of the ejaculate. Sections 1 = % motile cells, 2 = % live cells, and 3 = progressive motility scores.

prostatic versus vesicular fluids in each of the fractions. Most men are after instruction able to 'split' the ejaculate in four to six fractions.

It was recently reported that spermatozoa in the first parts of an ejaculate had a better motility and survival than those in the later parts. However, no difference was noticed between the spermatozoa in the first fractions and whole ejaculates from the same men (5). These first series of observations were made at a fixed time interval after ejaculation and subsequent investigations were therefore aiming at two other aspects: what changes could be recorded over a longer time period and to what extent was the changes in sperm survival and motility due to intra- and/or extracellular factors?

Figure 2A illustrates the changes in sperm motility (A-1: % motile cells), viability (A-2: % live cells), and progressive motility (A-3: progressive motility score – 0 = no progressive motility, 3 = good progressive motility) over a 4-hour period. Due to the time necessary for liquefaction 'zero' is actually 20–30 min after the ejaculation. From the figure it is obvious that spermatozoa ejaculated in the second half of an ejaculate have a significantly poorer survival and motility than those in the first portion. In another series of experiments (fig. 2B) albumin was added to the second portion immediately after the ejaculation (final concentration 4 %). This resulted in a marked improvement of sperm viability and no difference could now be recorded between cells in the two parts of the ejaculate (6).

The experiments reported above were performed with untreated samples. Due to the common distribution of spermatozoa the number of cells were significantly lower in the second part than in the first (cf. fig. 1). Spermatozoa were therefore pooled from several ejaculates and separated from the plasma by gentle centrifugation. They were then added to sperm-free seminal plasma collected by split ejaculation. In these experiments there were no differences between the survival of the cells in the first and last part of an ejaculate (7). This could mean that both intra- and extracellular factors were of importance for the first reported phenomenon or that the spermatozoa at the initial contact with the prostatic fluid had undergone certain changes, e.g. been coated with proteins that later protected them against any harmful effects excerted by the vesicular fluid.

This problem was approached by washing the spermatozoa carefully in a buffer solution containing albumin and then adding them to the various plasma fractions of split ejaculates (7). The results are partially presented in figure 3. Section A shows the changes in motility (A-1), viability (A-2), and progressive motility (A-3) with time in the first and second part of ejaculates, respectively. It is quite clear that the survival of the cells is markedly shortened in the vesicular fluid than in the prostatic. If, however, prostatic fluid had been added to the vesicular fluid in an amount equal to the original (fig. 3B) or albumin had been added to give a final concentration of 4 % (fig. 3C) the survival of the

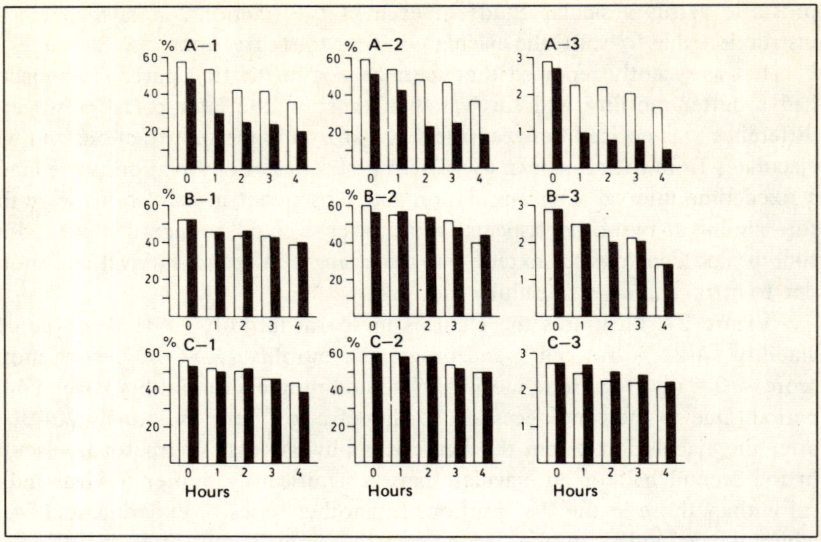

Fig. 3. Motility and survival of washed ejaculated human spermatozoa suspended in the first and second half of ejaculates. To the second half had been added an equal volume of 'prostatic' fluid (section B) or albumin (section C; final concentration 4 %). Section A = control. Sections 1–3 the same as in figure 2.

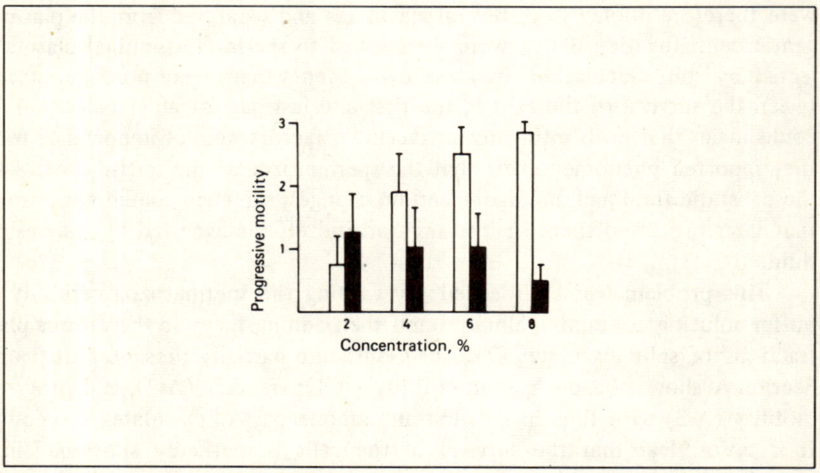

Fig. 4. Progressive motility score of washed ejaculated human spermatozoa suspended in a buffered salt solution after addition of increasing amounts of 'prostatic' (open bars) or 'vesicular' (closed bars) fluids.

spermatozoa was the same in the first and second fractions of the ejaculates. Careful washing of the spermatozoa thus can remove the factor that protects the spermatozoa from the negative influence of the vesicular fluid.

Spermatozoa that have been washed in an *albumin-free* buffer solution have many properties in common with epididymal spermatozoa, e.g. the same affinity to certain stains and the same motility pattern. We therefore call these 'epididymal-like' spermatozoa. The motility pattern is characterized by flagellating movement of the tails resulting in no or very weak progressive motility. If seminal plasma or albumin are added to a suspension of epididymal or washed ('epididymal-like') human spermatozoa progressive motility is immediately started. It was therefore of interest to study how fluid from the various fractions of split ejaculates influenced the motility pattern of 'epididymal-like' spermatozoa. In figure 4 the results from such a series of experiments are compiled. Only the results obtained with the first and last fractions (fraction 1 and 6, cf. fig. 1) of split ejaculates are included. At a 2-percent level the vesicular fluid gives a slightly better progressive motility than prostatic fluid but the progressive motility score increases proportionate to the concentration of the prostatic fluid and declines at higher concentrations of fluid from the seminal vesicles (8).

Our studies have thus demonstrated that the vesicular fluid contains one or more factors with a deleterious effect on sperm motility and survival. The prostatic fluid has, on the other hand, a marked stimulatory effect on the progressive motility of both washed ejaculated human spermatozoa and untreated epididymal human spermatozoa. In addition, as soon as the spermatozoa have been in contact with the prostatic fluid they are protected against the negative influence excerted by the vesicular fluid. The nature of the various factors involved is not yet clarified.

The observation that the motility, survival, and other functional parameters of the human spermatozoa are dependent on an optimal ratio between the prostatic and vesicular fluids is of interest from several points of view. It has been noted that the human seminal plasma has a chemical composition that for a given individual is not only remarkably constant but also independent of the ejaculatory frequency (9). This can only mean that the secretory activity in and the discharge of fluids from the accessory genital glands are under strict control. This, as well as the fact that the bulk of the spermatozoa appears in the first part of the ejaculate, can now be understood in terms of expediency. It also explains why many patients with biochemical signs of prostatic dysfunction have semen samples with poor sperm motility and viability, while patients with biochemical signs of an isolated vesicular dysfunction (low fructose) almost always have semen with good sperm motility and survival.

One practical aspect ought to be mentioned. In semen samples with poor sperm motility and survival and biochemical signs of a prostatic secretory dysfunction an improvement can be noted in sperm motility by adding albumin to

Table I. Effect of albumin on motility and viability of human spermatozoa

	Control time after ejaculation		Albumin time after ejaculation	
	½ h	4 h	½ h	4 h
Motile sperm, %	60	25	60	50
Live sperm, %	71	57	66	53
Progressive motility, score	2	1	3	3

In the first sample the motility was rapidly declining. The biochemical analyses indicated a prostatic secretory dysfunction. The second ejaculate was collected in a jar containing albumin (final conc. approximately 4 %). At this time the motility was initially better and maintained at a good level.

the sample immediately after the ejaculation. This is illustrated in table I. Similar effects on sperm motility have been reported after addition of caffeine to the semen (10, 11), but it is not likely that albumin and caffeine improve sperm motility through the same mechanisms.

It remains to be investigated if poor sperm viability in the seminal plasma is also correlated with a decreased survival in the female genital tract.

From the presentation above it could be questioned if the vesicular fluid only has negative effect in relation to sperm survival and transport. Such a question can at the present not be fully answered but it should be recalled that in man the seminal prostaglandins are secreted from the seminal vesicles (12). Addition of partially purified prostaglandins from the human seminal plasma (HSF-PG) or crystalline prostaglandins E_1 and $F_{2\alpha}$ have no effect on sperm motility or metabolism. However, prostaglandins are known to be extremely potent with regard to the uterine motility.

The possible significance of the prostaglandins (PG) in the human seminal plasma for human reproduction has challenged scientists since the observation by *Kurzrok and Lieb* (13) that very small amounts of human seminal fluid (HSF) could either decrease or increase the spontaneous motility of the isolated human myometrium.

Most tissues and body fluids contain one or more PG in amount from 1 ng to 1 or 2 mg/g wet weight. Human seminal plasma has 13 different PG and usually the total concentration is in the order of 80–120 μg/ml. In addition to the common PG (PGE_1, PGE_2, PGE_3, $PGF_{1\alpha}$, and $PGF_{2\alpha}$) it contains both PGA, PGB and the 19-hydroxylated forms of PGA and PGB.

With regard to the effects on the isolated human myometrium the various PG can be divided into two groups; the PGFs and the others. The F compounds cause increased motility of the isolated myometrium while the other PG inhibit the spontaneous motility of the nonpregnant myometrium and increase tonus and spontaneous activity of the pregnant uterus. Since the F compounds are present only in a few micrograms per ml the seminal plasma has an effect similar to that of PGE, PGA, and PGB – for references see *von Euler and Eliasson* (14).

From a physiological point of view it is interesting that intravaginal application of seminal fluid or partially purified prostaglandins from this fluid (HSF-PG) can influence the motility and tonus of both uterus and oviducts. This was first demonstrated by *Karlson* (15) and later by *Eliasson and Posse* (16). In these studies the uterine motility was recorded with intrauterine pressure receptors and the seminal fluid or HSF-PG was administered into the posterior part of vagina. Repeated experiments on the same women at different times during a menstrual cycle showed that a change in the uterine motility was recorded only at the estimated time of ovulation. The motility was increased but in some women this enhanced activity was followed by a period of decreased uterine motility (fig. 5). In two of these women the experiment was repeated the following period but this time an intravenous infusion of oxytocin-vasopressin was started a few minutes before application of HSF-PG. Prostaglandin under these circumstances only caused a marked decrease in uterine motility (fig. 6).

The reasons for giving oxytocin-vasopressin was twofold. *In vitro* it had been noted that sometimes a small dose of HSF-PG could cause an increase in the myometrial motility while a slightly larger dose caused the common relaxation. The reactivity pattern observed *in vivo* could thus be due to a gradual increase in PG-concentrations changing the motility from stimulation to inhibition. Oxytocin-vasopressin might change the reactivity of the uterus as suggested by *Csapo* (17) and thus make the myometrium more sensitive to the PG, i.e. the inhibition should be noted at a lower PG level in the tissue. On the other hand, later studies have not supported that this was the mode of action – for references see *von Euler and Eliasson* (14). The second reason to use a combination of oxytocin-vasopressin was that they probably are released during coitus and we wanted to study the effects of HSF-PG under as 'physiological' conditions as possible – for references see *Eliasson* (12) and *von Euler and Eliasson* (14).

Eliasson and Posse (18) also studied the effect of HSF-PG on the resistance to gas flow through the uterotubal junction and the Fallopian tubes (Rubin test). This investigation was performed on women with unexplained infertility.

In four out of seven patients intravaginal application of HSF-PG did not change the resistance to gas flow. In three women there was, on the other hand, a marked increase in the resistance. In two patients the gas flow had to be stopped – as exemplified in figure 7 – since the pain become too severe and the records indicated a complete occlusion at the utero-tubal junctions. Further studies are needed to clarify which of the two response patterns observed that should be regarded as 'normal'.

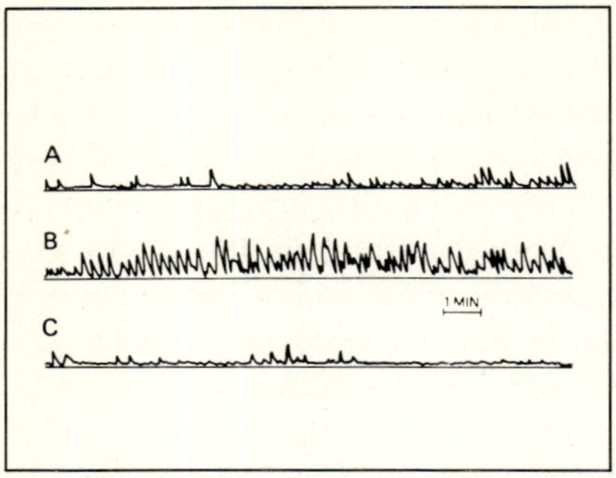

5

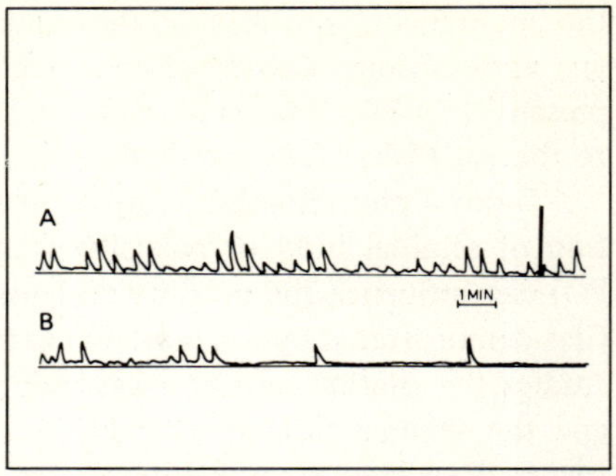

6

Fig. 5. Effect of PG on the motility of the human uterus *in vivo.* Registration from corpus at the time of ovulation. *A* Control period; *B* 5–20 min after intravaginal application of 150 U of PG; *C* 25–40 min after the application of PG. From *Eliasson* (4).

Fig. 6. Effect of PG on the motility of the human uterus *in vivo.* Registration from the corpus at the time of ovulation. During the experiment an i.v. infusion of 0.1 oxytoxin + 0.1 IU vasopressin is given. *A* Control period at single intravaginal application of 150 U of PG. *B* Records from 6 to 21 min after the application of PG. From *Eliasson* (4).

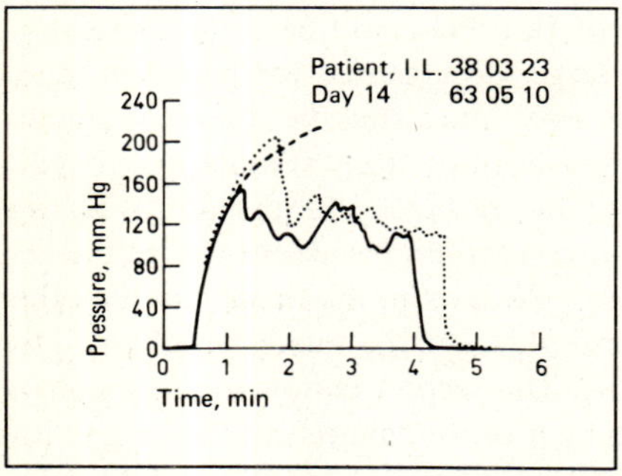

Fig. 7. Effect of PG on Rubin's test.——, control;········ and - - -, 8 and 25 or 40 min, respectively, after intravaginal application of 150 U of PG. From *Eliasson and Posse* (18).

In marked contrast to the results obtained in experiments with intravaginal application of HSF-PG are the results following intravenous infusions of PGE_1, PGE_2, and $PGF_{2\alpha}$. The pure compounds all cause an increase of tonus and of frequency and amplitude of the uterine contractions. The results are the same during different phases of the menstrual cycle and are not influenced by simultaneous infusion of oxytocin or vasopressin (19). The effects of intravenous injections of PGE_2 on the Fallopian tubes are different and only inhibition of tonus and/or contractions have been recorded. $PGF_{2\alpha}$ stimulates tubal motility (20).

Table II. Effects of partially purified PGs from human seminal fluid (HSF-PG), PGE_2 and $PGF_{2\alpha}$ on the motility/tonus of smooth muscles in nonpregnant human uterus and Fallopian tubes

	In vitro		*In situ*[1]	
	uterus	oviduct	uterus	oviduct
HSF-PG	–	–	+/–	+/0
PGE_2	–	–	+	–
$PGF_{2\alpha}$	+	+	+	+

+ = Stimulation; – = inhibition.
1 HSF-PG administered intravaginally; PGE_2 and $PGF_{2\alpha}$ intravenously.

The differences between the effects of intravaginal application of HSF-PG and intravenous infusion of pure PG on uterine and Fallopian tube motility are summarized in table II. The differences emphasize the importance to perform both physiological and pharmacological studies in our search for a role of PG in human reproduction.

During sexual stimulation there is a marked increase in the uterine activity and the intrauterine pressure. After the orgasm follows a period of low intrauterine tonus and motility (21). This reactivity pattern, the large amounts of PG normally present in the human seminal fluid and the recorded effects of PG on uterine motility indicate that these factors may play an important role in human reproduction.

It is not likely that a high PG content in the seminal plasma is a prerequisite for a passive sperm transport but PG may be one of many facilitating factors. If so, the seminal fluid contains factors that are of importance not only for the initiation of a progressive sperm motility but may also be vital for survival and migration of spermatozoa within the female genital tract.

References

1 *Hotchkiss, R.S.:* Fertility in men (Lippincott, Philadelphia 1944).

2 *Rozin, S.:* Studies on seminal plasma. Int. J. Fertil. *6:* 169–174 (1961).

3 *Eliasson, R.:* Correlation between the sperm density, morphology and motility and the secretory function of the accessory genital glands. Andrologie *2:* 165–169 (1970).

4 *Eliasson, R.:* Prostaglandin – properties, actions and significance. Biochem. Pharmacol. *12:* 405–412 (1963)

5 *Eliasson, R. and Lindholmer, C.:* Distribution and properties of spermatozoa in different fractions of split-ejaculates. Fertil. Steril. *23:* 252–256 (1972).

6 *Lindholmer, C. and Eliasson, R.:* The effect of albumin, magnesium, and zinc on

human sperm survival in different fractions of split ejaculates. Fertil. Steril. *25* (in press, 1974).

7 *Lindholmer, C.:* Survival of human spermatozoa in different fractions of split ejaculate. Fertil. Steril. *24:* 521–526 (1973).

8 *Lindholmer, C.:* The importance of seminal plasma for human sperm motility. Biol. Reprod. *10* (in press, 1974).

9 *Eliasson, R.:* Effect of frequent ejaculations on the composition of the human seminal plasma. J. Reprod. Fertil. *9:* 331–336 (1965).

10 *Haesungcharern, A. and Chulavatnatol, M.:* Stimulation of human spermatozoal motility by caffeine. Fertil. Steril. *24:* 662–665 (1973).

11 *Johnsen, Ø; Eliasson, R., and Abdel Kader, M.M.:* Effects of caffeine on the motility and metabolism of human spermatozoa. Andrologia *6:* 53–58 (1974).

12 *Eliasson, R.:* Studies on prostaglandin. Occurrence, formation and biological action; Diss., Stockholm. Acta physiol. scand. *46:* suppl. 158 (1959).

13 *Kurzrok, R. and Lieb, C.C.:* Biochemical studies of human semen. II. The action of semen on the human uterus. Proc. Soc. exp. Biol. Med. *28:* 268–272 (1930).

14 *Euler, U.S. von and Eliasson, R.:* Prostaglandins (Academic Press, New York 1967).

15 *Karlson, S.:* Spermas inverkan på motiliteten inom den icke gravida uterus hos människa. Nord. Med. *42:* 1466–1468 (1949).

16 *Eliasson, R. and Posse, N.:* The effect of prostaglandin on the nonpregnant human uterus *in vivo.* Acta obstet. gynecol. scand. *39:* 112–126 (1960).

17 *Csapo, A.:* Dependence of isometric tension and isotonic shortening of uterine muscle on temperature and on strength of stimulation. Amer. J. Physiol. *177:* 48–354 (1954).

18 *Eliasson, R. and Posse, N.:* Rubin's test before and after intravaginal application of prostaglandin. Int. J. Fertil. *10:* 373–377 (1965).

19 *Roth-Brandell, U.; Bygdeman, M., and Wiqvist, N.:* Effect of intravenous administration of prostaglandin E_1 and $F_{2\alpha}$ on the contractility of the nonpregnant human uterus *in vivo.* Acta obstet. gynec. scand. *49:* suppl. 5, pp. 19 (1970).

20 *Coutinho, E.M. and Maia, H.S.:* The contractile response of the human uterus, Fallopian tubes, and ovary to prostaglandins *in vivo.* Fertil. Steril. *22:* 539–543 (1971).

21 *Fox, C.A.:* Recent studies in human coital physiology. Clin. Endocrin. Metab. *2:* 527–543 (1973).

Authors' address: Prof. *R. Eliasson* and Dr. *C. Lindholmer,* Department of Physiology, Karolinska Institutet, Faculty of Medicine, *S–104 01 Stockholm* (Sweden)

Sperm Transport in Female Tract of Vertebrates

The Biology of Spermatozoa
INSERM Int. Symp., Nouzilly 1973, pp. 94–104 (Karger, Basel 1975)

Transport and Survival of Spermatozoa in Female Amphibia

Claude Boisseau and Jean Joly

Laboratoire de Biologie de la Reproduction[1], UER de Sciences Biologiques, Rennes

Introduction

Amphibia exhibit various physiological mechanisms to achieve fertilization. In anura, external fertilization is almost universal. On the other hand, in apoda and in caudata, except two primitive families, fertilization takes place internally, the question not yet being resolved for trachystomata. Examples of internal fertilization are known in oviparous as well as viviparous forms, in the aquatic environment as well as in animals adapted to terrestrial or subterranean courtship. In the absence of perfected copulatory organs, these amphibians show some adaptative specializations insuring insemination.

Spallanzani (45) was the first to demonstrate internal fertilization in newts during the courtship, without direct contact between the two sexes. But it was only a century later that the nature and the role of the spermatophore were understood (15, 38, 57).

In any female amphibians, the spermatozoa may be stored and their long-term survival is possible. *Leydig* (28) described in 1853 glands which open into the cloaca of female urodeles. Shortly later, *Siebold* (43) noted that these out-pocketings form a spermatheca that he named *receptaculum seminis,* thus opening the path for numerous subsequent studies (10, 20).

The important problem of sperm transport in the genital tract of female vertebrates has been recently reviewed (19, 47). Fertilization may be accomplished at different levels in the genital tract, raising several questions. In amphibians, these results remain fragmentary despite some recent findings.

1 Equipe de Recherche Associée au CNRS No. 129.

Sperm Transfer in Amphibians

Sperm Transfer in Anura and Apoda

External fertilization is the rule in most frogs and toads (1, 31, 37). There are several patterns of courtship, but in most instances the male, gripping the female with his front legs, fertilizes eggs during an axillary or lumbary amplexus. However, some species are characterized by internal fertilization. The torrent-living American species, *Ascaphus truei,* inseminates the female with an erectile intromissary organ kept in the female cloaca for 28–30 h (29, 31, 44). *Nectophrynoides occidentalis* (Bufonidae) is a small viviparous African toad with totally terrestrial habits (26). Related to its viviparity, insemination is effected by simple apposition of the cloacas.

In the oviparous (*Siphonops, Ichtyophis,* etc.) or viviparous (*Gymnopis, Typhlonectes,* etc.) apoda, fertilization is internal (31, 54, 55). The male possesses a large erectile copulatory organ which makes possible a copulation, the patterns of which have never been described. The males are provided with a developed Müllerian duct whose abundant secretions may supply a part of the fluid necessary for sperm transport (55).

Sperm Transfer in Caudata and Trachystomata

The mechanisms of sperm transfer in caudata have been extensively reviewed (7, 39, 49, 50). The main methods of insemination will be described here. In the two primitive families of the group, Hynobiidae (40, 48) and Cryptobranchidae (16), the eggs are fertilized after laying. The male *Ranodon sibiricus,* a high altitude Siberian hynobiid, deposits on the bed of the torrent in which it lives, a voluminous ribbon-shaped spermatophore against which the female applies its sacs of eggs (2, 8, 35). It seems to be an intermediary stage between external and internal fertilization (39, 49).

In higher forms of caudata, the eggs are fertilized within the body of the female. Typical courtship is unique among vertebrates and involves the deposition of a spermatophore on the substratum by the male, devoided of a true copulatory appendage. This gelatinous structure is surmounted by a small mass of spermatozoa (sperm cap) that the female takes up between her cloacal lips (15, 21, 31, 38–40, 42, 46, 57).

Females of trachystomata *(Pseudobranchus, Siren)* lacking spermatheca lay single eggs fixed to the substratum (13, 16, 31). It is not known whether fertilization is external or internal.

Sperm Survival in Anura and Apoda

Little is known about sperm survival in the two orders. Spermatozoa have been observed in the genital tracts of the female *Ascaphus truei* (29, 30). The gametes were found in the posterior part (or uterine segment) of the oviducts, in the lumen as well as the glands bordering the canal. After laying, numerous oviductal glands were still filled with sperm (30). In *Nectophrynoides,* ovulation generally occurs shortly after courtship. Sperm have not been recorded in the genital tract (*Xavier,* pers. comm.).

The females of gymnophiones (apoda) do not show any structure specialized for sperm storage (54, 55). Even though highly probable, at least at certain stages, the presence of spermatozoa in the genital tract does not seem to have been noted.

Sperm Storage in Caudata

In most urodeles, fertilization is internal. The sperm may be preserved for a variable time in the tubules which open at the level of the dorsal wall of the female cloaca. *Von Siebold* (43) referred to these tubules as 'receptaculum seminis' (5, 14, 20, 27, 57), or 'spermatheca' (5, 9, 10, 12, 36). For some authors (25), each individual tubule takes the name of spermatheca.

Anatomy and Histology of the Receptaculum Seminis

The spermatozoa are enclosed in glands which are homologous to the pelvic glands of the male cloaca (4, 27). The organization of the spermatheca varies according to the family (25). Thus, in the ambystomatidae (25) and salamandridae (14, 25, 27, 34), the tubules are numerous (up to 100) and open individually into the cloaca. On the contrary, in the plethodontidae (25), with the exception of *Hemidactylium scutatum* (12), the spermatheca is only made up of a few tubules linked with the cloaca by a common collecting canal.

The epithelium of the glands of the spermatheca is composed of secretory cells and a basal layer of flattened myoepithelial cells in all specimens studied. At the time of ovulation, the epithelium may be flattened, cubic, or even columnar as in *Salamandra salamandra* (fig. 2). In certain cases, it is thick but perforated by crypts of irregular form as in *Ambystoma* or *Euproctus* (11). In newts *(Triturus, Diemyctylus),* the epithelium is flattened and forms numerous processes in the lumen of the tubules (fig. 1). The connective tissue which enve-

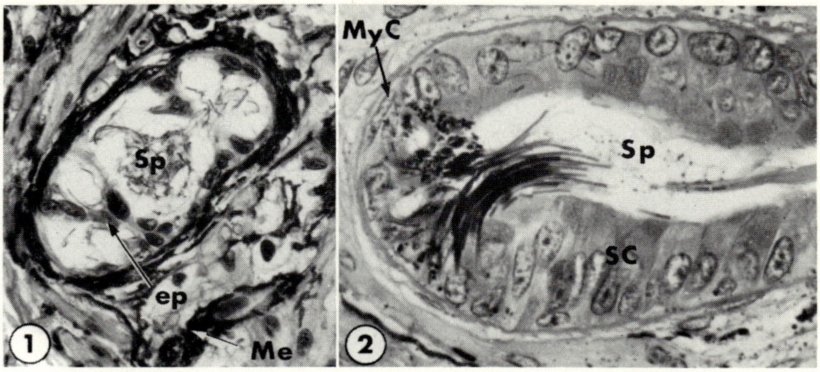

Fig. 1. Tubule of the spermatheca of the newt *Triturus helveticus* showing epithelial processes (ep). Me, melanophora; Sp, spermatozoa. × 243.

Fig. 2. Longitudinal section of tubule in *S. salamandra.* MyC, myoepithelial cell; SC, secretory cells, Sp, spermatozoa. Bouin's fluid, PAS. × 387.

lopes the tubules of the spermatheca encloses various cell types: smooth fibers (5), unmyelinated nerve fibers, fibroblasts, mast cells (10) and numerous melanophores, as has been noted by most authors.

Histochemistry and Ultrastructure of the Spermatheca

In some species, the spermatheca contains an abundant secretory product which reacts to PAS as well as to alcian blue at low pH – *Salamandra* (5, 20); *Pleurodeles* (27). In other species *(Taricha granulosa, Hydromantes genei),* the spermatheca is colored by PAS and not by alcian blue. Finally, in the genus *Triturus,* the secretory product is very scarce and only reacts feebly or not at all to PAS and to alcian blue. A certain heterogeneity is, therefore, observable despite the rather general presence of neutral or acid polysaccharides.

The ultrastructure of the secretory cells of the spermatheca has been the subject of recent studies (5, 36). In *Salamandra salamandra* (5), and *Notophthalmus viridescens* (10), the cells show a basal nucleus, a rather voluminous Golgi apparatus, and an apical mass of secretory granules. The granular reticulum is moderately developed and the apex of the cell shows numerous microvilli. Some glycogen is present in both species. In *Manculus quadridigittatus* (Plethodontidae), the voluminous Golgi apparatus develops dense granules. The mitochondria exhibit in this species a curious secretory activity (52).

The myoepithelial cells form an almost uninterrupted layer in contact with the basal lamina of the epithelium (5, 10). The myofilaments of these have the

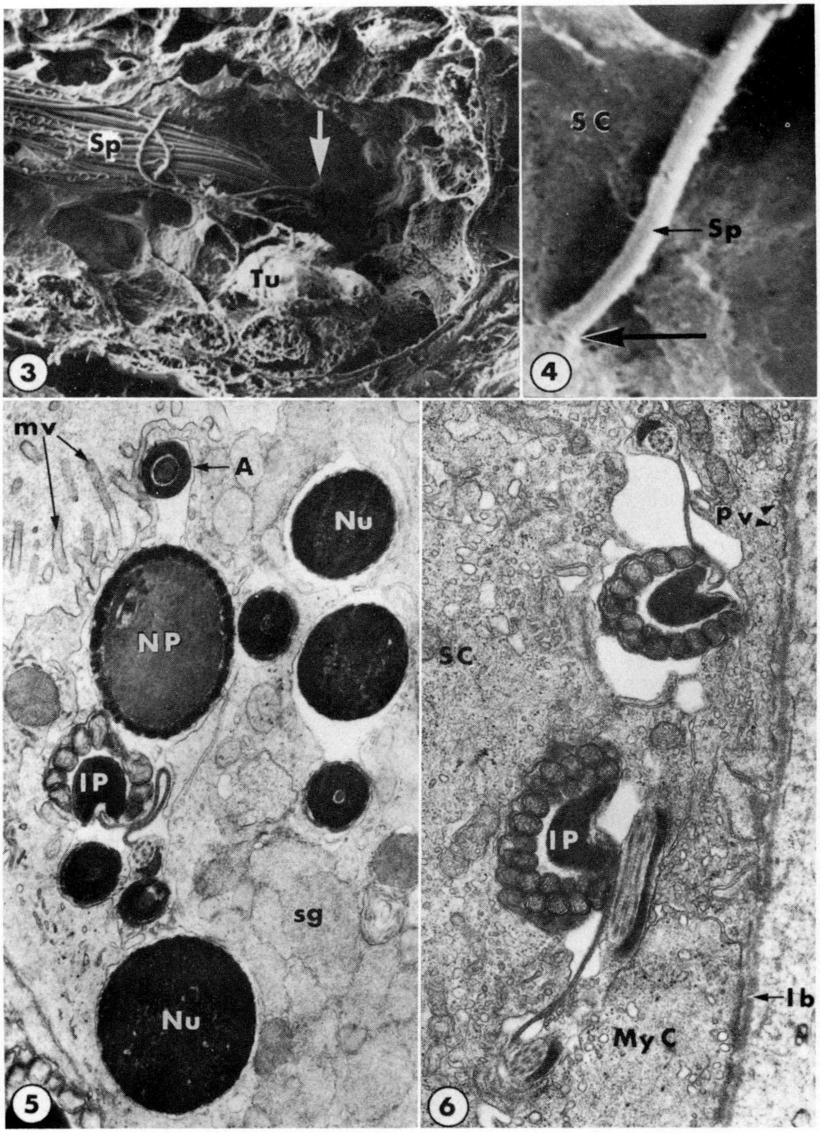

Fig. 3–6. Spermatheca of *S. salamandra.* Implantation (arrow) of a bundle of spermatozoa (Sp) in the bottom of a tubule (Tu). Scanning electron microscope. × 900. *4,* Detail of a spermatozoa (Sp) buried in a secretory cell (SC). × 3,600. *5,* Penetration of spermatozoa at the apex of a secretory cell. × 16,065. *6,* Deep penetration of spermatozoa in myoepithelial cells (MyC). × 14,850. A, Acrosome; Nu, nucleus; NP, neck piece; IP, intermediate piece of spermatozoa; lb, lamina basilis; mv, microvilli; pv, pinocytic vesicules; sg, secretory granules.

same diameter (75 Å) as the actin fibers of smooth muscle and show numerous contacts with the plasma membrane where pinocytosis is very active.

Physiology of the Spermatheca

Three phenomena are of special interest in spermathecal physiology: the penetration of sperm into the tubules; their survival in the spermatheca; and their evacuation for fertilization.

Penetration of Sperm in the Tubules of the Seminal Receptacles

The hypothesis of an autonomous migration of sperm has been generally accepted. Thus, *Noble and Weber* (33) observed the progression of sperm from the sperm-capsule of the spermatophore to the opening of the common duct of the spermatheca, into which they penetrate actively *(Desmognathus).* Similarly, the distribution in bundles of the sperms of *S. salamandra* (5, 20) supposes an active penetration of the gametes into the tubules (fig. 2, 3). The existence of a positive chemotropism of sperm for the products secreted by the epithelium is generally accepted (5, 20, 22, 27, 36), but has not been demonstrated experimentally, In *M. quadridigittatus, Pool and Hoage* (36) also suggest the possibility of sperm transport by the ciliated cells of the common collecting canal.

Duration of Survival and Relationships of Sperm with Spermatheca

Several observations testify to the survival of spermatozoa in the seminal receptacle and the persistence of their fertilizing capacity, since the period of ovulation often differs from that of insemination.

The survival time varies from a few weeks in most newts (27, 32) to more than two years in *S. salamandra* (3, 22). The sperms may be randomly distributed in the lumen of the tubules (27, 36) or in close contact with the epithelial processes (10). In *Diemyctilus pyrrhogaster* (51) and *S. salamandra* (20, 22, 34), they are deeply buried in the epithelium. The plasma membranes of the host cells are invaginated and form mantles surrounding the sperm (fig. 5, 6). Nearly all spermatozoa remain undamaged (5). It is possible that the proximity of the gametes to the cells of the seminal receptacle may have some nutritional significance to the sperm (5, 10, 20, 22) normally motionless in the tubules (5, 6). Glycogen (5, 10), such as mucoids secreted by the epithelium, may have some nutritive value. In *N. viridescens,* the presence of sperm in the tubules seems to increase the elaboration of secretory granules by the cells (10).

In the female *S. salamandra,* sperms are resorbed within several months after castration. The cells of the spermatheca, then, lose their secretory properties (22). Some experiments, using organ culture have shown in *S. salamandra* a direct role of the spermatheca in the survival of male gametes *in vitro* (5, 6).

Sperm Release

In the oviparous species, the compression of the spermatheca during the passage of eggs may contribute to the expulsion of spermatozoa. However, the role of myoepithelial cells (5, 10) and that of smooth fibers of the connective tissue (5, 10, 20) should not be overlooked. Moreover, the releasing of sperm in *S. salamandra* appears to be favored by the excretion of secretory products by the cells during ovulation (5).

Sperm Transport and Site of Fertilization in Female Amphibia

Except for *Ascaphus truei* (30), the presence of sperm has not yet been observed in the female genital tract of anura and apoda. However, in the latter, viviparity is not rare and it may be assumed that fertilization occurs in the oviducts (54). In the oviparous species of urodeles, spermatozoa are extruded onto the eggs passing near the spermatheca just before laying. The assertion (31), according to which fertilization occurs in the oviducts of newts, has been refuted by all subsequent data (7, 53).

The rare viviparous caudata, like *S. atra* and *S. salamandra* raise a more complex problem. In *S. atra,* a single egg, the 'embryonic' egg, develops in each uterus. The larva is fed during pregnancy by 20–40 'embryotrophic' eggs without jelly membrane. Contrary to an idea accepted for a long time (41, 52, 56) *Häfeli* (18) shows that only the 'embryonic' egg, coated with thick jelly layers, is fertilized in each uterus. This egg is expelled through the uterine orifice in the direction of the cloaca. A part of its external jelly layer then projects forming a mucilaginous plug which enters in contact with the spermatheca situated just ahead. Spermatozoa take this path to insure their fertilizing ability. No male gamete is seen in the others parts of the genital tract.

In *S. salamandra,* 10–40 larva develop in each uterus (23). The possible site of fertilization was discussed by many authors (17, 41, 53). Sperm ascent in the genital tract occurs at the time of ovulation (24). Spermatozoa can be observed in the caudal and middle oviduct, in the lumen and in the folds of the epithelium. Sperm heads may also be implanted in the glands of the oviduct. They are extremely rare in the cranial oviduct and absent from the ciliated tuba and from the uterus. After the transit of the eggs, the spermatozoa undergo phagocytosis. Thus, *S. salamandra* is, at present, the only caudata in which sperm ascent in the oviduct is demonstrated.

In *S. salamandra* or in *S. atra,* the ascent of sperm into the uterus is facilitated by the anatomical characteristics of the spermatheca, the tubules of which open at a very short distance from the uterine orifices. Muscular contraction is, therefore, sufficient to put the two organs in contact and to permit the passage

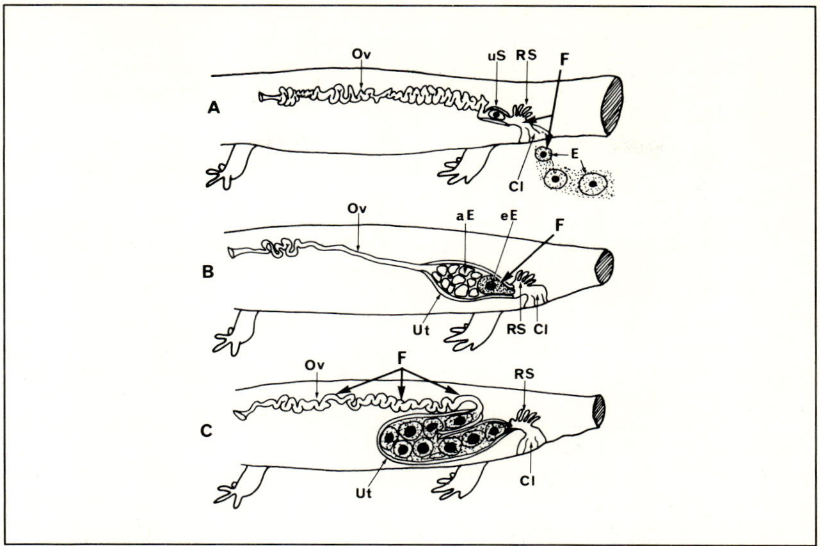

Fig. 7. Fertilization site in three urodeles. *A Pleurodeles waltlii,* oviparous species; fertilization (thick arrows) occurs in the cloaca just before oviposition. *B Salamandra atra,* viviparous species, in early pregnancy; fertilization of the 'embryonic' egg, occurs at the caudal extremity of the uterus (after *Häfeli,* modified). *C Salamandra salamandra,* in early pregnancy; fertilization have been performed during the passage of the eggs through the middle and caudal parts of the oviduct (see text). aE, Abortive eggs; Cl, cloacal cavity; E, laid eggs surrounded by jelly layers at various imbibition states; eE, 'embryonic' egg; F, site of fertilization; RS, receptaculum seminis; uS, uterine segment; Ut, uterus.

of the sperm. Mechanisms of sperm transport in the uterus and in the oviduct remain unknown. Figure 7 shows the main sites of fertilization in urodeles.

Future research is needed to delineate the physiological mechanisms of sperm transport in female amphibians.

References

1 *Angel, F.:* Vie et mœurs des amphibiens (Payot, Paris 1947).

2 *Bannikov, A.G.:* Die Biologie des Froschzahnmolches *Ranodon sibiricus* Kessler. Zool. Jb. Abt. Syst. Okol. Geogr. Tiere, Jena *86:* 245–252 (1958).

3 *Baylis, H.A.:* Delayed reproduction in the spotted salamander. Proc. zool. Soc., Lond. *109A:* 243–246 (1939).

4 *Blanchard, R.:* Sur les glandes cloacale et pelvienne et sur la papille cloacale des batraciens urodèles. Zool. Anz. *4:* 9–39 (1881).

5 *Boisseau, C.:* Cytologie ultrastructurale des organes (canal de Wolff, spermathèque)

assurant la conservation des spermatozoïdes chez *Salamandra salamandra* (L.). Etude de la survie spermatique en culture organotypique; Thèse, Rennes (1970).

6 *Boisseau, C.:* Etude en culture organotypique du rôle de la spermathèque et du canal de Wolff dans la survie des spermatozoïdes de *Salamandra salamandra* (L.). Ann. Sci. nat. zool. *13:* 163–180 (1971).

7 *Boisseau, C. and Joly, J.:* Transport et survie des spermatozoïdes chez la femelle des amphibiens. Symp. on Sperm Transport, Survival and Fertilizing Ability, Nouzilly 1973. INSERM *26:* 233–256 (1974).

8 *Briegleb, W.:* Die Spermatophore des Grottenolms. Zool. Anz. *166:* 87–91 (1961).

9 *Dawson, J.:* The cloaca and cloacal glands of Necturus. J. Morph. *36:* (1922).

10 *Dent, J.N.:* The ultrastructure of the spermatheca in the red spotted newt. J. Morph. *132:* 397–423 (1970).

11 *Despax, R.:* Contribution à l'étude anatomique et biologique des Batraciens urodèles du groupe des euproctes; Thèse, Paris (1923).

12 *Dieckman, J.M.:* The cloaca and spermatheca of *Hemidactylium scutatum.* Biol. Bull. *53:* 281–285 (1927).

13 *Dunn, E.R.:* The breeding habits of salamanders and their bearing on phylogeny. Copeia *115:* 25–28 (1923).

14 *Francis, E.T.B.:* The anatomy of the salamander (Clarendon Press, Oxford 1934).

15 *Gasco, F.:* Gli amori del Tritone alpestre *(Triton alpestris)* e la deposizione delle sue ova. Ann. Mus. Civ. Storia Nat. Genova *16:* (1880).

16 *Goin, C.J. and Goin, O.B.:* Introduction to herpetology (Freeman & Cy, San Francisco 1962).

17 *Grönroos, H.:* Zur Entwicklungsgeschichte des Erdsalamanders (*Salamandra maculosa* Laur.). Anat. H. *18:* 153–247 (1895).

18 *Häfeli, H.P.:* Zur Fortpflantzungsbiologie des Alpensalamanders (*Salamandra atra* Laur.). Rev. suisse Zool. *78:* 235–293 (1971).

19 *Hafez, E.S.E. and Thibault, C.* (ed.): Symp. on Sperm Transport, Survival and Fertilizing Ability, Nouzilly 1973. INSERM (1974).

20 *Joly, J.:* La conservation des spermatozoïdes et les particularités histophysiologiques du réceptacle séminal chez la salamandre *Salamandra salamandra* taeniata. C.R. Acad. Sci. *250:* 2269–2271 (1960).

21 *Joly, J.:* Sur l'éthologie sexuelle de *Salamandra salamandra* (L.). Z. Tierpsych. *23:* 8–27 (1966).

22 *Joly, J.:* Ecologie et cycles sexuels de *Salamandra salamandra* (L.): Thèse, CNRS (1966).

23 *Joly, J.:* Données écologiques sur la salamandre tachetée *Salamandra salamandra* (L.). Ann. Sci. nat. *10:* 301–366 (1968).

24 *Joly, J. et Boisseau, C.:* Localisation des spermatozoïdes dans l'oviducte de la salamandre terrestre, *Salamandra salamandra* (L.) (amphibien urodèle) au moment de la fécondation. C.R. Acad. Sci. *277:* 2537–2540 (1973).

25 *Kingsbury, B.F.:* The spermatheca and methods of fertilization of some American newts and salamanders. Proc. amer. Micros. Soc. *17:* 261–305 (1895).

26 *Lamotte, M.:* Observations écologiques sur les populations naturelles de *Nectophrynoides occidentalis* (fam. bufonidés). Bull. Biol. franç. Belge *93:* 355–413 (1959).

27 *Lemaitre-Lutz, F.:* Anatomie des glandes pelviennes de la femelle de *Pleurodèles waltlii* Michah.; leur rôle de réceptacle séminal. Ann. Embryol. Morph. *1:* 409–416 (1968).

28 *Leydig, F.:* Anatomisch-histologische Untersuchungen über Fische und Reptilien (Georg Reimer, Berlin 1853).

29 *Metter, D.E.:* On breeding and sperm retention in Ascaphus. Copeia *1964:* 710–711.

30 *Noble, G.K.:* Ontogeny and phylogeny within amphibia I. Amer. Mus. Novit. *165:* 1–17 (1925).

31 *Noble, G.K.:* The biology of the amphibia (Dover Publications, New York 1931).

32 *Noble, G.K. and Brady, M.K.:* Observations on the life history of the marbled salamander, *Ambystoma opacum* Gravenhorst. Zoologica, II *8:* 89–133 (1933).

33 *Noble, G.K. and Weber, J.A.:* The spermatophores of *Desmognathus* and other plethodontid salamanders. Amer. Mus. Novit. *351:* 1–15 (1929).

34 *Oppel, A. von:* Lehrbuch der vergleichenden mikroskopische Anatomie der Wirbelthiere, vol. 4 (G. Fischer, Jena 1904).

35 *Paraskiv, K.P.:* Semirechenskii Triton. Isvestija Akademii Nauk Kasakskoi SSR *8:* 47–56 (1953).

36 *Pool, T.B. and Hoage, T.R.:* The ultrastructure of secretion in the spermatheca of the salamander *Manculus quadridigittatus* (Hollbrook). Tissue Cell *5:* 303–313 (1973).

37 *Rabb, G.B. and Rabb, M.S.:* Additional observations on breeding behavior of the surinam toad, *Pipa pipa.* Copeia *4:* 636–642 (1963).

38 *Robin, C.:* Observations sur la fécondation des urodèles. J. Anat. Physiol. *10:* 376–390 (1874).

39 *Salthe, S.N.:* Courtship patterns and the phylogeny of the urodeles. Copeia *1967:* 100–117.

40 *Sasaki, M.:* On a Japanese salamander in lake Kuttarush which propagates like the axolotl. J. Coll. Agr. Hokkaido Imp. Univ. *15:* 12–23 (1924).

41 *Schwalbe, G.:* Zur Biologie und Entwicklungsgeschicht von *Salamandra atra* und *maculosa.* Z. Biol. *34:* 340–396 (1897).

42 *Shoop, R.:* Aspects of reproduction in Louisiana Necturus populations. Amer. Mid. Nat. *74:* 357–367 (1965).

43 *Siebold, C.T. von:* Über das Receptaculum seminis der weiblichen Urodelen. Z. wiss. Zool. *9:* 463–495 (1858).

44 *Slater, J.R.:* The mating of *Ascaphus truei* Stejneger. Copeia *1931:* 62–63.

45 *Spallanzani, L.:* Expériences pour servir à l'histoire de la génération des animaux et des plantes; avec une ébauche de l'histoire des êtres organisés avant leur fécondation (J. Senebier, Genève 1785).

46 *Stebbins, R.C.:* Natural history of the salamanders of the plethodontid genus *Ensatina.* Univ. Calif. Publ. Zool. *54:* 47–124 (1954).

47 *Thibault, C.G.:* Sperm transport and storage in vertebrates. J. Reprod. Fertil. *18:* 39–53 (1973).

48 *Thorn R.:* Contribution à l'étude d'une salamandre japonaise l'*Hynobius nebulosus* (Schlegel). Comportement et reproduction en captivité. Arch. Inst. Grand-Duc. Luxembourg Sci. nat. *29:* 210–215 (1963).

49 *Thorn, R.:* Salamandres et tritons. Aperçu sur l'ordre des urodèles avec quelques considérations sur la phylogénie de ces animaux. Bull. Soc. Nat. Luxembourg *69:* 101–111 (1966).

50 *Thorn, R.:* Les salamandres d'Europe, d'Asie et d'Afrique du Nord (Lechevalier, Paris 1968).

51 *Tsutsui, Y.:* Notes on the behavior of the common Japanese newt *Diemyctylus pyrrhogaster* Boie. I. Breeding habits. Mem. Coll. Sci. Kyoto Imp. Univ. *7:* 159–167 (1931).

52 *Vilter A. et Vilter, V.:* Rôle de l'oviducts dans l'uniparité utérine de la salamandre noire des Alpes orientales (*Salamandra atra* Laur.). C.R. Soc. Biol. *156:* 49–51 (1962).

53 *Wahlert, G. von:* Eileiter, Laich und Kloake der Salamandriden. Zool. Jb. Anat. *73:* 276–324 (1953).

54 *Wake, M.H.:* Evolutionary morphology of the caecilian urogenital system. I. The gonads and the fat bodies. J. Morph. *126:* 291–331 (1968).

55 *Wake, M.H.:* Evolutionary morphology of the caecilian urogenital system. II. The kidneys and urogenital ducts. Acta anat. *75:* 321–358 (1970).

56 *Weber, A.:* La fécondation chez la salamandre alpestre (*Salamandra atra* Laur.). C.R. Ass. Anat. *17:* 327–329 (1922).

57 *Zeller, E. von:* Untersuchungen über die Samenträger und den Kloakemvulst der Tritonen. Z. wiss. Zool. *74:* 171–221 (1905).

Authors' address: *C. Boisseau* and Prof. *J. Joly,* Biologie de la Reproduction, Université de Rennes, Avenue du Général Leclerc, *F–35031 Rennes Cédex* (France)

The Biology of Spermatozoa
INSERM Int. Symp., Nouzilly 1973, pp. 105–113 (Karger, Basel 1975)

Sperm Survival and Transport in the Female Genital Tract of Reptiles

H. Saint-Girons

Laboratoire d'Evolution des Etres Organisés, Paris

Observation of animals in captivity long ago showed the possibility of long spermatozoa survival in the female genital tract of reptiles, e.g. tortoises (13, 14, 24), snakes (2, 6, 21, 23, 26, 29, 40–42) or lizards (1).

The site of storage of the spermatozoa and their progression through the female genital tract were studied much later and in a small number of squamates. However, these rare cases demonstrate the large variety of mechanisms to insure fertilization of the ova at various stages of the sexual cycle. Unfortunately, to our knowledge, there is no exact observation in crocodiles.

Structure of the Female Genital Tract in Lepidosauria and Chelonia

To understand the way in which the spermatozoa sometimes survive for long periods in the female genital tract of reptiles, it is necessary to describe briefly the structure of the oviduct. From a sinus of complex structure which varies from one species to another (20), there successively develop a vaginal segment without glands, then the uterus characterized by the presence of shell-secreting glands, the tuba or Fallopian tube where the albumin-secreting glands are situated (when present), and the infundibulum.

In its caudal part, the vaginal segment is a cylindrical, narrow, convoluted tube (the vaginal tube). On a cross section, the epithelium has large longitudinal folds in its lumen which have a triangular tongue shape converging towards the center. In its cranial part, the vaginal segment becomes enlarged and flattened transversely. This area, which we call the utero-vaginal area, is very small in the sphenodon and in tortoises, but in numerous lizards, it corresponds to a large, flattened expanded area which becomes moulded on the coprodaeum. The epithelium of the vaginal segment consists of two cell types, usually in equal number and regularly alternating: the ciliated cells and the mucous cells secreting a mucopolysaccharide acid. The vaginal segment of the infra-order Iguania has seminal receptacles in the form of simple crypts varying in depth in the utero-vaginal area, or in the form of tubes which penetrate into the chorion of the vaginal finger-like projections (8, 17, 33).

The uterine glands occupy the major part of the chorion in the uterus, which represents the major part of the oviduct and the site where the embryos of ovoviviparous species develop. The epithelium has few folds and consists of ciliated cells and secretory cells, the latter being more abundant. They present the same histochemical characteristics as similar cells found in the vaginal tube, but there is generally very little mucin.

The oviduct forms tight S-shaped folds. This segment is particularly well developed in tortoises where the albumin-producing glands are very numerous. In the Lepidosauria these glands are lacking or very rare. In all cases, the tuba is covered with an epithelium consisting of high prismatic mucous cells containing a large quantity of very acid mucin, and also of large ciliated cells which are 3 or 4 times less numerous than the former cells. At the base of the epithelial folds in squamates, there are often ciliated crypts. It is at this level in snakes that the tubular seminal receptacles discovered by *Fox* (16) in 1956 become differentiated. The end of the tube (infundibulum) is a thin covering without any glandular cells which does not seem to have any role in the survival or transport of spermatozoa.

Sexual Cycles and Duration of Sperm Storage

In all reptiles, the spermatozoa are deposited in the cloaca at the cranial extremity of the genital sinus, and immediately after copulation reach the vaginal segment. But the timing of the latter phenomenon varies considerably from one species to another.

In several lizards, copulation often precedes ovulation by only one or two weeks (29). The spermatozoa first remain within the vaginal segment then, a few days before ovulation, they reach the tuba where fertilization occurs. When copulation takes place earlier, the spermatozoa remain for a longer period in the vaginal segment. This probably is what happens in *Hemiergis peronii,* the females of which copulate in autumn and lay their eggs in spring (39). Usually, there are no longer any spermatozoa in the female genital tract after laying and further copulation is necessary before the following ovulation. However, there are some cases in the Iguania suborder, in which isolated females have given rise to several successive fertilized eggs (1, 9, 33). This possibility seems to be due to the previously described seminal receptacles within the vaginal segment.

In most, if not in all snakes, copulation precedes ovulation by at least one month and often much more. There spermatozoa, first stored within the lumen of the vaginal tube, reach the tuba 2–6 weeks before ovulation, and many become stored within the tubular seminal receptacles (12, 16, 27, 28, 30, 32, 38). When copulation occurs in the autumn, the spermatozoa may remain within the vaginal segment for 7–8 months; the duration of the period within the tuba is unchanged. On histological examination, no spermatozoa have ever been found within the oviduct of females autopsied more than two months after ovulation. However, there are undoubted cases of isolated females having laid fertilized eggs on successive occasions (10, 22, 23, 26, 42). Unfortunately, it is not known whether in these cases the spermatozoa remain stored within the tubular seminal receptacles or whether they return to the vaginal segment between ovulation and egg laying.

According to *Dawbin* (11), about ten months may elapse between copulation and the laying of the only living Rhynchocephalia, *Sphenodon punctatus,* an oviparous species. As there are apparently no specialized seminal receptacles (19), one may suppose that the spermatozoa are stored for most of this time within the lumen of the vaginal tube.

Whereas the tortoises are the first certain example of delayed fertilization in reptiles (24), we still do not know where the spermatozoa are stored. In *Emys orbicularis* (29) and *Clemnys leprosa* (7), copulation occurs over a period of 2–3 months preceding ovulation. In females of neither species, autopsied after laying their eggs, did we find any spermatozoa or specialized seminal receptacles in the oviduct.

Structure and Role of the Seminal Receptacles in Reptiles

Seminal Receptacles of the Vaginal Segment

In numerous cases and especially in snakes, the spermatozoa are simply stored for a few weeks or months within the lumen or folds of the vaginal segment. They may remain in direct contact with the epithelium or be regularly placed along the latter and seemingly fixed by their achrosome to the apex of the mucous cells. In various Iguanidae (8) and Agamidae (35), the spermatozoa are grouped together within more or less deep crypts of the utero-vaginal area, the covering epithelium of which does not differ from the surrounding epithelium. The spermatozoa are generally fixed to the epithelial cells by their achrosome and, in certain cases, the whole head is engaged within the epithelium which then loses its ciliated border (8). In at least one species, *Uta stansburiana* (9), isolated females laid fertilized eggs on two successive occasions, which shows that simple epithelial crypts may insure the protection of at least some spermatozoa during the passage of the eggs. However, when they laid eggs a third time, these were always sterile. Finally, in chameleons (33) and *Anolis carolinensis*

(17), the seminal receptacles in the form of long finger-shaped tubes are more specialized and penetrate into the chorion of the vaginal tube. These structures are lined with cubic cells which have neither cilia nor secretory structures. The spermatozoa are stored there in small groups within the axis of the tube, the head sometimes in contact with the epithelial cells, but never fixed to the latter. Under these conditions, spermatozoa may insure the fertilization of at least three successive batches of eggs or more in the absence of males.

Seminal Receptacles in the Tuba

The tuba of all reptiles include numerous epithelial folds where the spermatozoa may be lodged for a short time before ovulation. In rare Chelonia studied from this point of view (personal observation), *Sphenodon punctatus* (19) and various Sauria and Amphisbenia (5, 37), there was no particular structure in the tuba except that ciliated cells may become more numerous in the depths of the folds. The development of true ciliated crypts, grouped together in small clusters, was however observed in 3 Gekkonidae, 2 Iguanidae, 1 Agamidae, 1 Scincidae and 1 Anguidae (8, 34, 35). The tubal seminal receptacles are not well differentiated except in snakes, where their presence seems to be constant (15, 18, 35). Their development during the sexual cycle has been followed in a few species (12, 16, 30–32, 38) and *Hoffman and Wimsatt* (25) have studied them under the electron microscope. During periods of sexual rest, the deeply ramified invaginations in the epithelium, which constitute the tubal seminal receptacles are almost entirely covered with ciliated cells and very small nonciliated ovoid or cubic cells having no secretion may only be distinguished in their depths by careful examination. These structures become gradually developed in spring. They contain glycogen as ciliated cells elsewhere, and lipid inclusions and traces of acid mucin. One month before ovulation, vacuoles appear which finally occupy a large part of the cell. When the spermatozoa reach the tuba (within 2–6 weeks before ovulation), they become rapidly concentrated within the seminal receptacles. Many remain in the ciliated part of the tubes but many also penetrate into the deep narrow part and seem to penetrate the epithelial cells in small groups. Electron microscopy reveals no true penetration within the cell as the various membranes remain intact. Furthermore, there is no difference between spermatozoa which have remained within the seminal receptacles and others; this is hardly evidence in favor of maturation at this level. The involution of the tubal seminal receptacles starts immediately after ovulation and is practically over within one month. During this period, small groups of spermatozoa may be found at the bottom of the tubes which have regressed somewhat but which have always entirely disappeared from the female genital tract before the end of the summer in the four species studied.

Under these conditions, female snakes may lay fertilized eggs on several successive occasions when isolated from males. The fact that the percentage of fertilized eggs generally falls, tends to prove that this is due to exhaustion of a reserve of spermatozoa, and not due to parthenogenesis. However, the sperm survival mechanism is unknown.

Discussion

The fact that in numerous species, the spermatozoa may survive within the lumen of the vaginal segment, casts some doubt on the role played by the seminal receptacles of this area or the close relation of spermatozoa and epithelial cells in the nourishment of sperm. The essential role of the seminal receptacles of the vaginal segment seems to permit survival of at least some quantity of spermatozoa during passage of the eggs. Perhaps when the receptacles consist of simple crypts, the attachment of the spermatozoa to the epithelial cells favors this survival. Unfortunately, it is not known whether this attachment alone, which occurs in certain lizards such as *Xantusia henshawi* and Gekkonidae, can compensate for the absence of any seminal receptacle. In most reptile species, all mature females are fertilized between egg laying and the following ovulation. However, in special circumstances such as scarcity of males in certain populations, or difficulty for members of the opposite sex to meet in very sparse populations, the ability of the female to lay several fertile eggs after only one insemination definitely contributes to survival of the species, and in other cases, may allow some expansion of the population per year.

The functional role of the tubal seminal receptacles is not clear. There is no index of maturation of the spermatozoa at this level and many of them remain in the ciliated part of the crypts. This also suggests that they may be nourished when not in direct contact with the cells in the bottom of the tubes. However, the tubular seminal receptacles, like the corresponding receptacles in the vaginal segment, permit preservation of a certain number of spermatozoa after passage of the eggs. These spermatozoa are no doubt available for fertilization of later eggs in certain isolated females.

Sperm Migration

The movements of spermatozoa in the reptile female genital tract are only known from histological examination of the oviducts removed at various stages of the sexual cycle. There are two periods of sperm migration. The first is immediately after copulation when spermatozoa reach the vaginal segment. Apparently most of them rapidly leave the genital sinus, and as the latter is often lacking in ciliated cells, this movement is probably active, and oriented by a concentration gradient of oviduct secretions. The spermatozoa become distributed in a variable manner once in the vaginal segment, depending on the species, when they seem to lose their motility.

The second and most important migration leads the spermatozoa from the vaginal segment to the tuba. As the oviduct is entirely covered with ciliated cells, this migration may be passive. However, histological appearances are in favor of active movement. The spermatozoa are almost always found in the center of the lumen, in the uterus, and are not in contact with the epithelial surface. Naturally, this is only a presumption as artifacts are always possible during fixation. In any case, this migration is definitely as rapid and massive, if judged by the difficulty of observation, and by the fact that when the spermatozoa appear in the tuba, there no longer remain any spermatozoa within the lumen of the vaginal segment.

The migration of the spermatozoa towards the tube coincides in time with numerous endocrine changes which affect the oviduct (4, 34, 36). For example, this is the moment when the mucous cells of the tuba which have a different secretion from that of the vaginal segment, start to excrete into the lumen. This secretion may reach the caudal part of the oviduct and stimulate the spermatozoa which are situated there. In Iguanidae, *Cuellar* (8) considers the possibility of a stimulation of the seminal receptacles by muscle contractions. From an anatomical point of view, there is nothing against this theory. However, muscle contractions alone cannot expel the spermatozoa from the epithelial crypts, and even less from the finger-shaped tubes. Chemical stimulation may be necessary to explain migration of the spermatozoa towards the tuba at an exact stage of the sexual cycle.

Expulsion of the spermatozoa from the seminal receptacles of the tuba may probably be explained, as suggested by *Fox* (16), by the mechanical effect of the passage of the ovules. However, *Hoffman and Wimsatt* (25) consider further explanations such as the role of intense vacuolization of the cells in the bottom of the tubes, or increased osmotic pressure due to breakdown of intracellular mucopolysaccharides.

Conclusion

Our knowledge of survival and movements of spermatozoa within the female genital tract of reptiles is still incomplete. However, the following conclusions may be provisionally drawn:

1. The storage of spermatozoa, even when prolonged for several months, may occur in the lumen of the vaginal segment without close contact with the epithelial cells.

2. When they exist, the seminal receptacles of the vaginal segment insure mechanical protection of the spermatozoa during passage of the eggs above all.

3. Little is known about the functional role of the seminal receptacles of the tuba, which are well developed only in snakes.

4. Migration of the spermatozoa from the vaginal segment towards the tuba seems to be rapid and active. The nature of the releasing mechanism is unknown.

The possibility of long spermatozoon survival in the female genital tract is, together with a latency period in embryonic development (3), one of the mechanisms which renders the time of birth of the young independent of the sexual cycle of the male. This has some ecological significance. Furthermore, the possibility of sperm survival from one laying to another maintains normal reproductive levels when, for environmental or physiological reasons, the chances of regular insemination of females are reduced.

References

1 *Atsatt, S.R.:* Storage of sperm in the female chameleon *Microsaura pumila pumila.* Copeia *59* (1953).

2 *Blanchard, F.N. and Blanchard, F.C.:* The inherance of melanism in the garter snake, *Thamnophis sirtalis sirtalis* (Linnaeus), and some evidence of effective autumn mating. Mich. Acad. Sci. *26:* 177–193 (1940).

3 *Bons, J. et Bons, N.:* Notes sur la reproduction et le développement de *Chamaeleo chamaelo* (L.). Bull. Soc. Sci. nat., Maroc. *40:* 323–335 (1960).

4 *Bons-Guenier, N.:* Etude du cycle sexuel d'*Acanthodactylus erythrurus lineomaculatus* Dum. et Bibron (Sauria, Lacertidae); Thèse, Montpellier (1966).

5 *Bons, J. et Saint-Girons, H.:* Ecologie et cycle sexuel des Amphisbéniens du Maroc. Bull. Soc. Sci. nat., Maroc. *43:* 117–158 (1963).

6 *Carson, H.L.:* Delayed fertilization in a captive indigo snake with notes on feeding and shedding. Copeia *1945:* 222–225.

7 *Combescot, C.:* Sexualité et cycle génital de la Tortue d'eau algérienne, *Emys leprosa* Schw. Bull. Soc. Hist. nat. Afr. Nord *45:* 366–377 (1954).

8 *Cuellar, O.:* Oviductal anatomy and sperm storage structures in lizards. J. Morph. *119:* 7–20 (1966).

9 *Cuellar, O.:* Delayed fertilization in the lizard *Uta stansburiana.* Copeia *1966:* 549–552.

10 *Darevsky, L.S.:* Delayed fertilization in the Brazilian colubrid snake, *Xenodon merremii* (Wagler). J. Herpet. *5:* 82–83 (1971).

11 *Dawbin, W.H.:* The tuatara in its natural habitat. Endeavour *21:* 16–24 (1962).

12 *Duguy, R. et Saint-Girons, H.:* Cycle annuel d'activité et reproduction de la Couleuvre vipérine, *Natrix maura* (L.). Terre et Vie *1966:* 423–457.

13 *Ewing, H.E.:* Reproduction in the eastern box turtle, *Terrapene carolina carolina* (Linné). Copeia *1933:* 95–96.

14 *Ewing, H.E.:* Continued fertility in female box-turtle following mating. Copeia *1943:* 112–114.

15 *Fawcett, J.D.; Stavitsky, A.H., and Smith, H.M.:* Oviductal anatomy and sperm storage

in *Tantilla nigriceps, T. atriceps,* and *T. gracilis* (Serpentes, Colubridae). J. Colo.-Wyo. Acad. Sci. *7:* 76 (1972).

16 *Fox, W.:* Seminal receptacles of snakes. Anat. Rec. *124:* 519–540 (1956).

17 *Fox, W.:* Special tubules for sperm storage in female lizards. Nature, Lond. *198:* 500–501 (1963).

18 *Fox, W. and Dessauer, H.C.:* The single right oviduct and other urogenital structures of female *Typhlops* and *Leptotyphlops.* Copeia *1962:* 590–597.

19 *Gabe, M. et Saint-Girons, H.:* Histologie de *Sphenodon punctatus* (Centre national de la Recherche scientifique, Paris 1964).

20 *Gabe, M. et Saint-Girons, H.:* Contribution à la morphologie comparée du cloaque et des glandes épidermoïdes de la région cloacale chez les Lépidosauriens. Mém. Mus. nat. Hist. nat., Paris *33:* 149–292 (1965).

21 *Gloyd, H.K.:* Studies on the breeding habits and young of the copperhead *Agkistrodon mokasen* Beauvois. Mich. Acad. Sci. *19:* 587–604 (1933).

22 *Groves, J.D.:* Delayed fertilization in the snake *Boiga dendrophila.* Herpetologica *29:* 20–22 (1973).

23 *Haines, T.P.:* Delayed fertilization in *Leptodeira annulata polysticta.* Copeia *1940:* 116–118.

24 *Hildebrandt, S.F.:* Review of experiments on artificial culture of diamondback terrapin. Bull. US Bureau Fish. *45:* 25–70 (1929).

25 *Hoffman, L.H. and Wimsatt, W.A.:* Histochemical and electron microscopic observations on the sperm receptacles in the garter snake oviduct. Amer. J. Anat. *134:* 71–96 (1972).

26 *Kopstein, F.:* Ein Beitrag zu Eierkunde und zur Fortpflanzung der Malaiischen Reptilien. Bull. Raffles Mus., Singapore *14:* 81–167 (1938).

27 *Ludwig, M. and Rahn, H.:* Sperm storage and copulatory adjustment in the prairie rattle snake. Copeia *1943:* 15–18.

28 *Rahn, H.:* Sperm viability in the uterus of the garter snake, Thamnophis. Copeia *1940:* 109–115.

29 *Rollinat, R.:* La vie des reptiles de la France centrale (Delagrave, Paris 1934).

30 *Saint-Girons, H.:* Le cycle sexuel chez *Vipera aspis* dans l'ouest de la France. Bull. Biol. Franç. Belg. *91:* 284–350 (1957).

31 *Saint-Girons, H.:* Données histochimiques sur les glucides de l'appareil génital chez les vipères, au cours du cycle reproducteur. Ann. Histochim. *4:* 235–243 (1959).

32 *Saint-Girons, H.:* Le cycle reproducteur de la vipère à cornes, *Cerastes cerastes* (L.), dans la nature et en captivité. Bull. Zool. France *87:* 41–51 (1962).

33 *Saint-Girons, H.:* Présence de réceptacles séminaux chez les caméléons. Beaufortia *9:* 165–172 (1962).

34 *Saint-Girons, H.:* Données histophysiologiques sur le cycle annuel des glandes endocrines et de leurs effecteurs chez l'Orvet, *Anguis fragilis* (L.). Arch. Anat. Micr. *52:* 1–51 (1963).

35 *Saint-Girons, H.:* Déplacements et survie des spermatozoïdes chez les Reptiles. Coll. sur Transport, Survie et Pouvoir fécondant des Spermatozoïdes, Nouzilly. INSERM *26:* 259–281 (1973).

36 *Saint-Girons, H. et Duguy, R.:* Données histophysiologiques sur le cycle annuel de l'hypophyse chez *Vipera aspis* (L.). Z. Zellforsch. *56:* 819–853 (1962).

37 *Saint-Girons, H. et Duguy, R.:* Le cycle sexuel de *Lacerta muralis* (L.) en plaine et en montagne. Bull. Mus. nat. Hist. nat. *42:* 609–625 (1970).

38 *Saint-Girons, H. et Kramer, E.:* Le cycle sexuel chez *Vipera berus* (L.) en montagne. Rev. suisse Zool. *70:* 191–221 (1963).

39 *Smyth, M. and Smith, M.J.:* Obligatory sperm storage in the skink *Hemiergis peronii.* Science *161:* 575–576 (1968).

40 *Trapido, H.:* Mating time and sperm viability in Storeria. Copeia *1940:* 107–109.

41 *Wall, F.:* A popular treatise on the common Indian snakes. I. J. Bombay Nat. Hist. Soc. *16:* 533–554 (1905).

42 *Woodward, S.F.:* A few notes on the persistance of active spermatozoa in the African night adder, *Causus rhombeatus.* Proc. zool. Soc., Lond. *103:* 189–190 (1933).

Author's address: *H. Saint-Girons,* Laboratoire d'Evolution des Etres Organisés, 105, bd Raspail, *F–75006 Paris* (France)

The Biology of Spermatozoa
INSERM Int. Symp., Nouzilly 1973, pp. 114–122 (Karger, Basel 1975)

Maturation and Transport of Avian Spermatozoa

G.A.R. Kamar and E.S.E. Hafez

Faculty of Agriculture, Cairo University, Giza, and Wayne State University School of Medicine, Detroit, Mich.

The long duration of fertilizing ability of avian sperm in the female reproductive tract remained obscure until *Van Drimmelen* (24) revealed some of the physiological characteristics of the oviduct and sperm involved in these processes.

Sperm Transport

Sperm are rapidly distributed throughout the oviduct after AI or copulation, but they disappear from the lumen by 24 h (table I).

After Natural Mating

Copulation enhances sperm distribution in the oviduct (24), and the rate of cilia beats (2).

After AI

The UVJ appears to be a barrier to sperm movement up the oviduct; it prevents transport of dead sperm, and permits transport of live sperms when deposited intravaginally. Massive numbers of sperm engorge the infundibulum following intra-magnal or infundibular AI. After intrauterine AI, considerable numbers of sperm are found in the uterus, while only moderate numbers enter the infundibulum after 10–15 min. When semen is deposited anterior to UVJ, large numbers of sperm reach the infundibulum. Dense sperm concentrations are found in the uterus, while few are found in the magnum 45 min after intravaginal AI is performed, or 15 min after oviposition (3). The efficiency of sperm

Table I. The time required for sperm to reach the uterovaginal junction (UVJ) and infundibulum in chickens

Techniques of introducing sperm	Time spent to reach		References
	UVJ, min	infundibulum, min	
Natural mating	–	15	(1)
Natural mating	–	60	(15)
AI in uterus	–	26	(14)
AI in uterus	–	15	(15)
AI in vagina	45	3–15	(3)
AI in vagina	15	15	(16)

transport of semen deposited above UVJ is greater than that deposited below it. Radioactive live and dead turkey sperm inseminated intravaginally reach the infundibulum within 15 min of insemination (9). Live sperm reached the uterus and infundibulum in greater numbers than did dead sperm. The majority of both radioactive live (58 %) and dead (76 %) sperm, however, were excreted as early as 15 min following insemination. The majority of sperm seem to be concentrated in the infundibulum and the uterus.

Effect of Oviductal Milieu

Oviposition

The presence of hard-shelled eggs in the uterus has no effect on transport of live sperm to the infundibulum, while the presence of a yolk and albumen in the magnum is a barrier to sperm transport. After the disappearance of sperm from the oviductal lumen, they reappear only about the time of ovulation or laying (2). Washing the oviduct with saline does not cause the release of sperm, however, passing an egg in the oviduct squeezes and stretches the glands containing the sperm and releases them. Inhibition of ovulation and oviposition by hormones does not prevent the movement of sperm past the UVJ, but it slows this movement and the rate at which it empties from the UVJ glands. Time of oviposition does not affect sperm distribution after intravaginal insemination of labelled semen with ^{32}P (4). Also, sperm release is not confined to any type of ovulatory cycle. Oviposition and ovulation affect sperm distribution in the oviduct only by mechanical means pertaining to egg or ovum passage, not to a hormonal or cyclic rhythm in the oviduct. Also, at each oviposition, large numbers of sperm are expelled outside the cloaca. Some sperm are also diffused in the yolk vitelline membrane and the albumin.

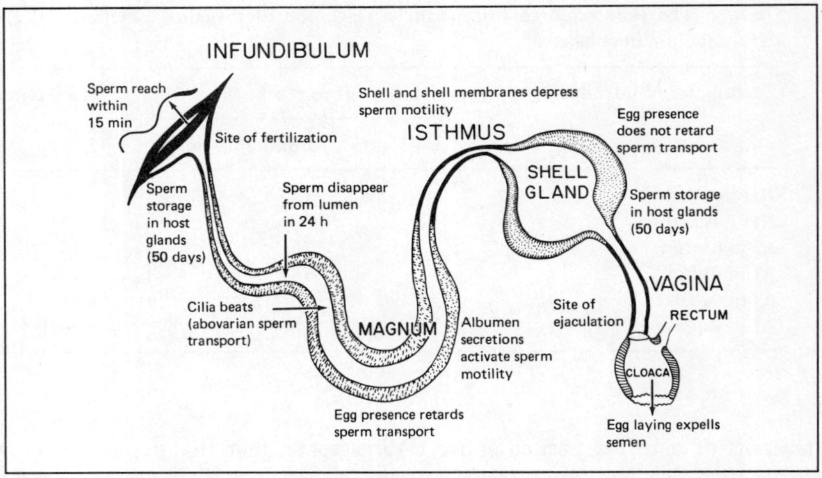

Fig. 1. Diagramatic illustration of the patterns of transport of avain spermatozoa in the female reproductive tract.

Physiological Mechanism

Abovarian sperm transport takes place by either muscular activity of the oviduct, intrinsic sperm motility, ciliary movement of the oviductal epithelium or by all these means. UVJ may also play a role in this respect. Dead sperm deposited in front of the UVJ or transported anteriorly is as efficient and rapid as live sperm; however, when dead sperm are deposited behind the UVJ none can be detected in the infundibulum. It seems that the cilia posterior to the UVJ beats the sperm towards the cloaca so that most of the semen deposited in the vagina (90 %) leaks out because of this movement (3). The remainder of the sperm, by their intrinsic motility, can enter the UVJ and remain there or travel by other means to the infundibulum. There are two types of oviductal ciliary movements in birds, adovarian and abovarian; they occur prior to ejaculation then are reactivated prior to fertilization. The sperm somehow are protected from premature aging and retain their capacity for fertilization even for days or weeks. These mechanisms are illustrated in figure 1.

Effect on Sperm Morphology and Motility

Sperm recovered by washing the oviduct are morphologically normal and motile, though percentage of motility ranges from 50 to 100 %. All the sperm found in the sperm host glands are morphologically normal. This observation suggests that these glands filter out abnormal sperm. Semen volume within certain limits (from 0.1 to 0.25 ml) has little or no effect on the distribution of sperm in the oviduct lumen, or the concentration or density of sperm in sperm host glands in the UVJ and infundibulum region. However, when semen volume is reduced, sperm distribution is adversely affected. The oviduct was ligated at several regions and inseminated at each portion. In the infundibulum, many motile sperm were found until the 21st day following insemination. In the uterus, sperm were also found until the 21st day. But less motile sperm were found than infundibulum. However, there were no motile sperm in smears taken after the 3rd day following insemination in the other regions of the oviduct (21). *In vivo* sperm in albumin and infundibular secretions become immotile at body temperature, but remained motile in uterine secretions (23). Motile sperm can be maintained *in vitro* for at least two days in fresh infundibular tissue in a plasma medium (22). The biochemical environment in infundibulum glands is quite different from UVJ glands. The infundibular glands supply the sperm with less nutritive substances than the UVJ (13). Accordingly, the sperm survive longer in the host glands of UVJ than in the infundibular glands.

Sperm Storage

Nature of Residence

Location

Early studies indicate that the storage site of sperm was in the infundibulum or the so-called sperm host gland. Later, it was found that the main host glands located in UVJ play the major part in fertilization. As the infundibulum was the normal fertilizing site of the eggs, the sperm were released from the UVJ and transported to the infundibulum region via the oviduct. This suggests that the majority of sperm are stored in the UVJ glands and the minority in the mucosal folds of the chalaziferous region (infundibulum). These two sperm storage sites are subsequently termed 'sperm host glands' (fig. 1). The migration of sperm into the UVJ glands occurs by the motility of sperm. Sperm were found in smears collected from each oviductal part until the fourth day following insemination. Thereafter, the presence of sperm was limited to the lower region of infundibulum and UVJ. In smears collected from UVJ, they were found until the 22nd day (16).

Number of Sperm Host Glands

The number of host glands detected in either normal residence regions of the oviduct varied greatly according to the physiological status of the oviduct, insemination techniques, and sperm concentration in the ejaculate. No consideration has been given to intra- and interbird variation.

Anatomical Aspects

The fimbriae is interrupted with shallow pits of grooves lined with nonciliated glandular-type cells between the major folds. The transition between funnel and chalaziferous region is manifested by a gradual transformation of grooves into tubular glands. Ultrastructure studies of the UVJ glands indicate that they are of a tubular type and their cells are secretory, as evidenced by swollen microvilli on the apical surface and large droplets of secretory material in the gland lumena. These glands have greatly enlarged lumen. However, there is no evidence of changes in cellular activity related to the egg-laying cycle of hens sacrificed at various stages of the ovulatory cycle. Histochemical tests of the UVJ of hens demonstrate that host glands contain a lipid in granules which is presumably secretory. They also secrete acid phosphatase (18). Lipids then react at least to some phospholipid reagents; however, they have abnormal solubility properties and differ histochemically in the apical and basal portions of the gland cells. Intracellular cholestrol ester can be detected. All these nutritive materials aid in the prolongation of the life span of sperm stored in the oviductal glands either in the infundibulum or the UVJ (6). Electron microscopical studies have revealed that the acrosomes tend to be imbedded among microvilli of the gland cells (26). Structural deterioration of the sperm occurs after extended periods of storage (16).

Effect of AI

The lumen of the infundibulum, and the mucosa of the chalaziferous region in the infundibulum, are the only regions where sperm are found in the normal and motile state for a period of 3–14 days following vaginal and intraperitonial AI. Few sperm are found in the cavity of the lower region of infundibulum until the 15th day following AI.

Sperm are found in every UVJ gland after intravaginal insemination. Meanwhile, sperm do not penetrate the infundibular glands until 2 h or more after intravaginal insemination. Within a day sperm disappear in washings from any part of the oviduct, except the infundibulum. Most sperm deposited in the infundibulum are preserved in the host glands of this portion, or may migrate and enter the UVJ gland. Most of sperm deposited in the vagina preserved in the UVJ glands.

Sperm introduced via laparatomy migrate only anteriorly and remain in the infundibulum following the passage of ova. Turkey semen was inseminated in chickens and histological examinations were done in the infundibulum and UVJ. Sperm were found in UVJ glands only after intravaginal inseminations. After intramagnal insemination, sperm were present almost exclusively in the infundibulum; and after intrauterine insemination sperm accumulated in both regions (18).

Sperm Survival and Duration of Fertilizing Ability

The time between insemination and the laying of the last fertile egg ranges from 9 to 21 days with an average of 11 days. Over 90 % of the eggs laid on the 6th day following insemination are fertile; then tend to gradually decline.

Effect of Semen Characteristics

Semen and sperm characteristics are so closely interrelated that it is difficult to study them separately. Though they are useful for predicting fertility from a mating or an AI, they are not helpful in explaining the underlying causes of poor fecundity in semen. Probably, each abnormality is only one manifestation of a more general deficiency. The characteristics that correlate with poor fertility are: deficient sperm concentration and hydrogenase activity (as measured by methylene blue or resazurin reduction time) excessive number of 'dead' sperm, and certain other morphological abnormalities. Usually, poor-quality semen accounts for a shortened fertility period. Low fertility may be associated with inability of the sperm to enter the sperm host glands.

Fertility Duration

Lacking survival of sperm within the lumen of the oviduct is probably brief, similar to that in mammals. The fluids in the lumen may be hostile to sperm compared to those of glands. However, sperm survive within the avian female for long periods: up to 32 days in the chicken, and 70 days in the turkey, though the commonly known 'fertile period' is usually shorter. Individuals and breeds vary considerably (table II).

Sperm Concentration

Reduction in numbers of sperm at the anterior end of the oviduct leads to reduced fecundity. On the other hand, there is evidence suggesting that an excessive number of sperm at the infundibulum leads to early disturbances of embryonic development. The minimum number of sperm which have to be inseminated to ensure fertility is proportional to the efficiency of sperm transport, which is greatly influenced by the site of insemination.

Fertility is high for the first week following insemination of more than 150 million sperm, but it decreases rapidly when sperm are decreased to 30 million. No fertile eggs are obtained by insemination of 4 million (20). After the intro-

Table II. Duration of egg fertility of different species

Species	Time fertilizing capacity retained, days	References
Goose *(Anser anser)*		(10)
Mallard duck *(Anas platyrhynchos)*	7–10	(5)
Ringdove *(Zenaidura macroura)*		(17)
Bobwhite quail *(Coturnix coturnix)*		(11)
Chicken *(Gallus gallus)*	10–14	
Ring-necked pheasant *(Phasians colchicus)*		(19)
Turkey *(Melagris gallopawo)*	22–46	(27)

duction of 1, 10, 100, and 1,000 million sperm into the infundibulum, egg fertility percentages in the first week are 51, 78, 90, and 100, respectively. Egg fertility duration is 9, 13, 19 and 19 days, respectively. Following intravaginal insemination with the same numbers of sperm, lower corresponding figures are observed, 21, 28, 95 and 97 %, and fertility duration is 4, 7, 14, and 16 days (20, 21).

In over 100 million sperm, there were no differences in the percentage of fertile eggs nor in the duration of fertilities following vaginal inseminations (20, 21).

AI Techniques

Vaginal insemination just after oviposition results in complete infertility of ova ovulated within 23 min after the insemination: while it fertilizes about 75 % of ova shed the same day of the insemination (3). Long fertility duration (21–26 days) follows the deposition of 0.05 ml of semen in the anterior oviduct, compared to a shorter duration (11–16 days) following midvaginal insemination with the same dose (12).

Fertility for 2 weeks, or the duration of fertility following a single intravaginal insemination with 100 million sperm, is higher and longer than that following a single infundibular insemination with ten million sperm. Duration of fertility following the infundibular insemination is longer (12–27 days) than the intravaginal insemination (6–17 days). The use of intrauterine insemination may increase the number of functionally intact sperm reaching the infundibulum, thus increasing the fertility level and its duration. Intravaginal insemination shows low viability and fertilizing ability (7). In both groups fertility during the first week was similar (25). Intramagnal insemination, however, adversely affects egg production and causes exceptionally high incidence of pre-ovipositmal embryonic mortality, when compared to results from insemination by techniques which deposited sperm in the UVJ region (26).

Intramagnal insemination causes a high incidence of early embryonic mortality, whereas insemination of germ-free semen from vas deferens results in reduced embryonic mortality.

References

1 *Allen, T.E. and Grigg, G.W.:* Sperm transport in the fowl. Austr. J. Agric. Res. *8:* 788–799 (1957).

2 *Bobr, L.W.; Lorenz, F.W., and Ogasawara, F.X.:* Distribution of spermatozoa in the oviduct and fertility in domestic birds. I. Residence sites of spermatozoa in fowl oviducts. J. Reprod. Fertil. *8:* 39–47 (1964).

3 *Bobr, L.W.; Lorenz, F.W., and Ogasawara, F.X.:* Distribution of spermatozoa in the oviduct and fertility in domestic birds. II. Transport of spermatozoa in the fowl oviduct. J. Reprod. Fertil. *8:* 49–58 (1964).

4 *Burke, W.H.; Ogasawara, F.X., and Fuqua, C.L.:* Transport of spermatozoa to the site of fertilization in the absence of oviposition and ovulation in the chicken. Poultry Sci. *48:* 602–608 (1969).

5 *Elder, W.H. and Weller, M.W.:* Duration of fertility in the domestic mallard hen after isolation from the drake. J. Wildlife Mgmt *18:* 495–502 (1954).

6 *Fujii, S.:* Histological and histochemical studies on the oviduct of the domestic fowl with special reference to the region of utero-vaginal juncture. Arch. histol. jap. *23:* 447–459 (1963).

7 *Grigg, G.W.:* Efficiency of sperm transport and fertility. Nature, Lond. *201:* 570–571 (1964).

8 *Gilbert, A.B.:* Formation of the egg in the domestic chicken, in *McLaren* Advances in reproductive physiology, vol. 2 (Academic Press, New York 1966).

9 *Howarth, B., jr.:* Transport of spermatozoa in the reproductive tract of turkey hens. Poultry Sci. *50:* 84–89 (1971).

10 *Johnson, A.S.:* Artificial insemination and duration of fertility of geese. Poultry Sci. *33:* 638–640 (1954).

11 *Kulenkamp, A.W.; Coleman, T.H., and Ernst, R.A.:* Artificial insemination of bobwhite quail. Poultry Sci. *44:* 1392 (1965).

12 *Lorenz, F.W. and Ogasawara, F.X.:* Distribution of spermatozoa in the oviduct and fertility in domestic birds. VI. The relations of fertility and embryo normality with site of experimental insemination. J. Reprod. Fertil. *16:* 445–455 (1968).

13 *Lorenz, F.W.; Reynolds, M.E., and Gilbert, A.B.:* Distribution of spermatozoa in the oviduct of domestic birds. Physiologist *10:* 236 (1967).

14 *Mimura, H.:* On the mechanism of travel of spermatozoa through the oviduct in the domestic fowl, with special reference to the artificial insemination. Okajimas folia anat. jap. *17:* 459–476 (1939).

15 *Parker, J.E.:* Reproductive physiology in poultry; in *Hafez* Reproduction in farm animals (Lea & Febiger, Philadelphia 1968).

16 *Prasad, S.:* Histological location and role of uterovaginal glands in storage and release of spermatozoa in chickens and turkeys. Diss. Abstr. B *28:* 2114-B (1967).

17 *Riddle, O. and Behre, E.H.:* Studies on the physiology of reproduction of birds. IX. On the relation of stale sperm to fertility and sex in ring doves. Amer. J. Physiol. *57:* 228–249 (1921).

18 *Schindler, H.; Kempenich-Pinto, O.; Broda, H., and Ben-David, E.:* Storage and survival of turkey spermatozoa in the domestic hen's oviduct. J. Reprod. Fertil. *21:* 359–361 (1970).

19 *Shick, C.:* Sex ratio-egg fertility relationships in the ring-necked pheasants. J. Wildlife Mgmt *11:* 302–306 (1947).

20 *Takeda, A.:* Behavior of spermatozoa in the genital tract of the hen. I. Presence of spermatozoa in the oviduct. Jap. Poultry Sci. *1:* 19–31 (1964).

21 *Takeda, A.:* Behavior of spermatozoa in the genital tract of the hen. III. Fertility and spermatozoal behavior following infundibulal insemination. Jap. Poultry Sci. *3:* 15–22 (1966).

22 *Tanaka, K. and Koga, O.:* Studies on the maintenance of fertilizing spermatozoa in the hen's oviduct. Jap. J. zootech. Sci. *42:* 237–249 (1971).

23 *Tingari, M.D.:* The fine structure of the epithelial lining of the excurrent duct system of the testis of the domestic fowl *Gallus domesticus.* Quart. J. exp. Physiol. *57:* 271–278 (1972).

24 *Drimmelen, G.C. van:* 'Spermnests' in the oviduct of the domestic hen. J. Sth afr. vet. med. Ass. *17:* 42–45 (1946).

25 *Krey, H.P. van:* Storage and transport of spermatozoa within the oviduct of the domestic fowl. Diss. abstr. *26:* Diss. No. 65-9296 (1965).

26 *Krey, H.P. van; Ogasawara, F.X., and Lorenz, F.W.:* Distribution of spermatozoa in the oviduct and fertility in domestic birds. IV. Fertility of spermatozoa from infundibular and uterovaginal glands. J. Reprod. Fertil. *11:* 257–262 (1966).

27 *Tienhoven, A. van and Steel, R.G.D.:* The effect of different dilutents and dilation rates on fertilizing capacity of turkey semen. Poultry Sci. *36:* 473–479 (1957).

Authors' addresses: Prof. *G.A.R. Kamar,* Animal Science Department, Faculty of Agriculture, Cairo University, *Giza* (Egypt); Prof. *E.S.E. Hafez,* Department of Gynecology and Obstetrics, C.S. Mott Center for Human Growth and Development, Wayne State University School of Medicine, *Detroit, MI 48201* (USA)

The Biology of Spermatozoa
INSERM Int. Symp., Nouzilly 1973, pp. 123–133 (Karger, Basel 1975)

The Relationship between Stored Spermatozoa and the Oviducal Epithelium in Bats of the Genus *Tylonycteris*[1]

P.A. Racey, Fumie Suzuki and Lord Medway

Unit of Reproductive Biology, University of Liverpool, Liverpool, and Great Glemham House, Saxmundham

I. Introduction

That spermatozoa remain fertile after prolonged storage in the reproductive tract has been established in vespertilionid bats of North America and Europe (11). In the species concerned, mating is initiated in autumn and ovulation and fertilization occur in spring, so that part of the period of sperm storage coincides with hibernation, when body temperatures fall to low levels. Sperm storage has also been found to occur at tropical latitudes in the vespertilionids *Pipistrellus ceylonicus* in India (2) and *Tylonycteris robustula* and *T. pachypus* in Peninsular Malaysia (7). The interval between copulation and ovulation is about six weeks in *P. ceylonicus* and up to two months in *Tylonycteris.*

The sites of sperm storage vary among bats. In *P. pipistrellus* the uterus is distended with spermatozoa (13) but in *P. ceylonicus* sperm are stored in both the uterus and oviducts (2). In *Tylonycteris,* comparatively few sperm are found in the uterus and most occur in the oviduct where they are orientated with their heads towards the epithelium (7). The present paper reports a further investigation of the relationship between spermatozoa and their storage organs in the bats of this genus.

1 Travel and collection expenses were met by a grant from the Royal Society to Lord *Medway. P.A. Racey* was supported by the Ford Foundation and *F. Suzuki* by the Population Council.

II. Materials and Methods

17 female ***Tylonycteris pachypus*** and seven female ***T. robustula*** were removed from their characteristic roosts in the internodal spaces of bamboo (9) in the vicinity of the University of Malaya Field Studies Centre, Ulu Gombak (3 ° 20′N; 101 ° 45′E) between 29th November and 11th December, 1972, when, according to the schedule established in previous years, all were expected to be in oestrus and to have mated.

Standard procedures were used to prepare the reproductive tracts for light and electron microscopy. These, together with the histochemical tests applied, are detailed by *Racey et al.* (14).

III. Observations

A. Site of Sperm Storage

The uterus of *Tylonycteris* is bicornuate. In freshly killed specimens, the distal part of the oviduct appeared full of opaque white material which formed a tapered streak, flocculated in appearance, and slightly reflective and opalescent. Histological examination revealed that this material consisted of spermatozoa, which were concentrated throughout the junctura and isthmus regions of the oviduct with a few in the ampullary region (fig. 1). Sperm seldom appeared free in the lumen of the oviduct, and almost all were distributed in a single layer adjacent to the epithelium and with their heads orientated towards it. Few sperm were found in the uterus of any of the specimens examined; when present they were situated in the distal part of the horn, around the uterotubal junction.

B. Relationship between Spermatozoa and Oviducal Epithelium

The heads of a few spermatozoa were embedded either in the microvilli or cilia which originate from the cells of the oviducal epithelium. The heads of most spermatozoa were situated, however, in indentations in the luminal surface of the epithelial cells (fig. 2, 3). These indentations vary in depth from those which enclose only the acrosome of the spermatozoa to those which enclose the whole head.

When the indentations are examined at higher magnification, it can be seen that the plasma membrane around the head of the sperm is closely applied to the

---

Fig. 1. Sperm orientated towards epithelial cells of the oviducal gland in the junctura region of the oviduct of *Tylonycteris pachypus.* × 770.

Fig. 2. Electron micrograph showing spermatozoa in the isthmus region of the oviduct of *Tylonycteris pachypus.* Ciliated cells are abundant in this region and sperm heads are embedded in cilia as well as in indentations in epithelial cells. × 7,500.

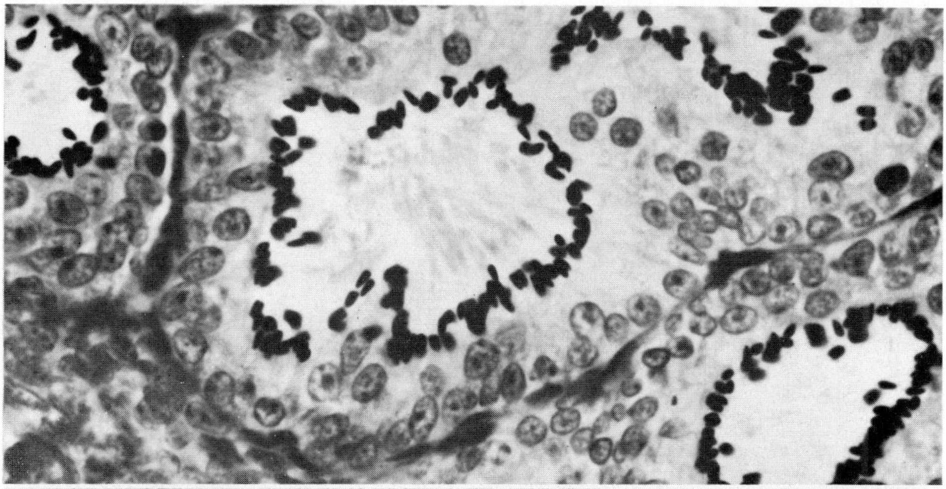

1

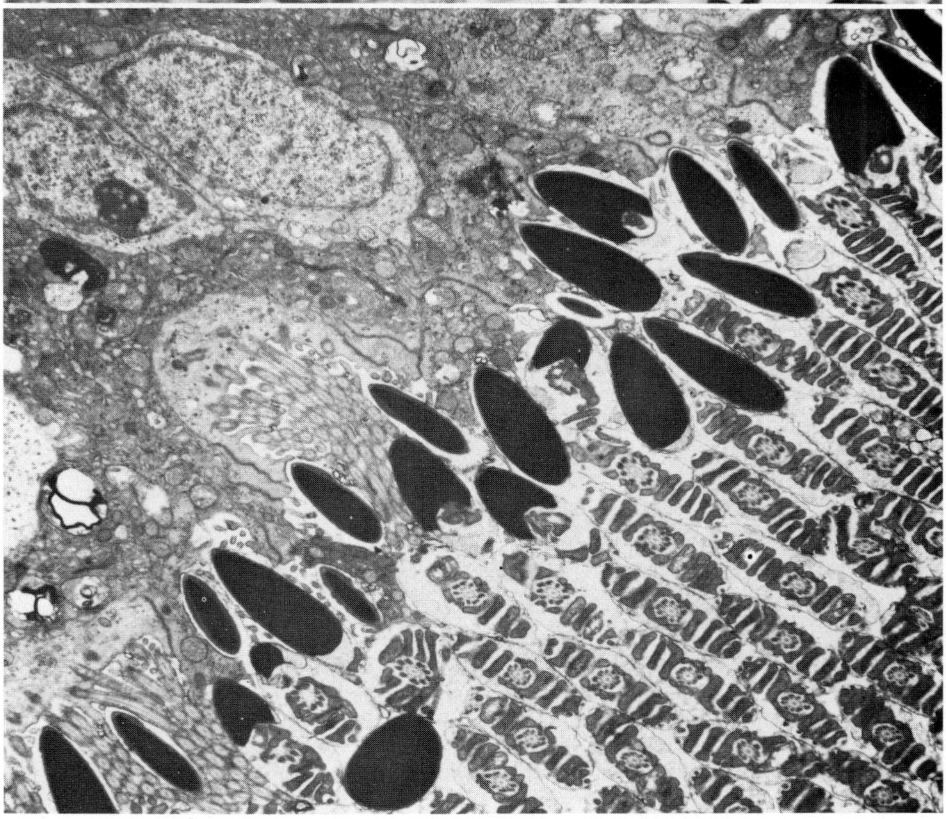

2

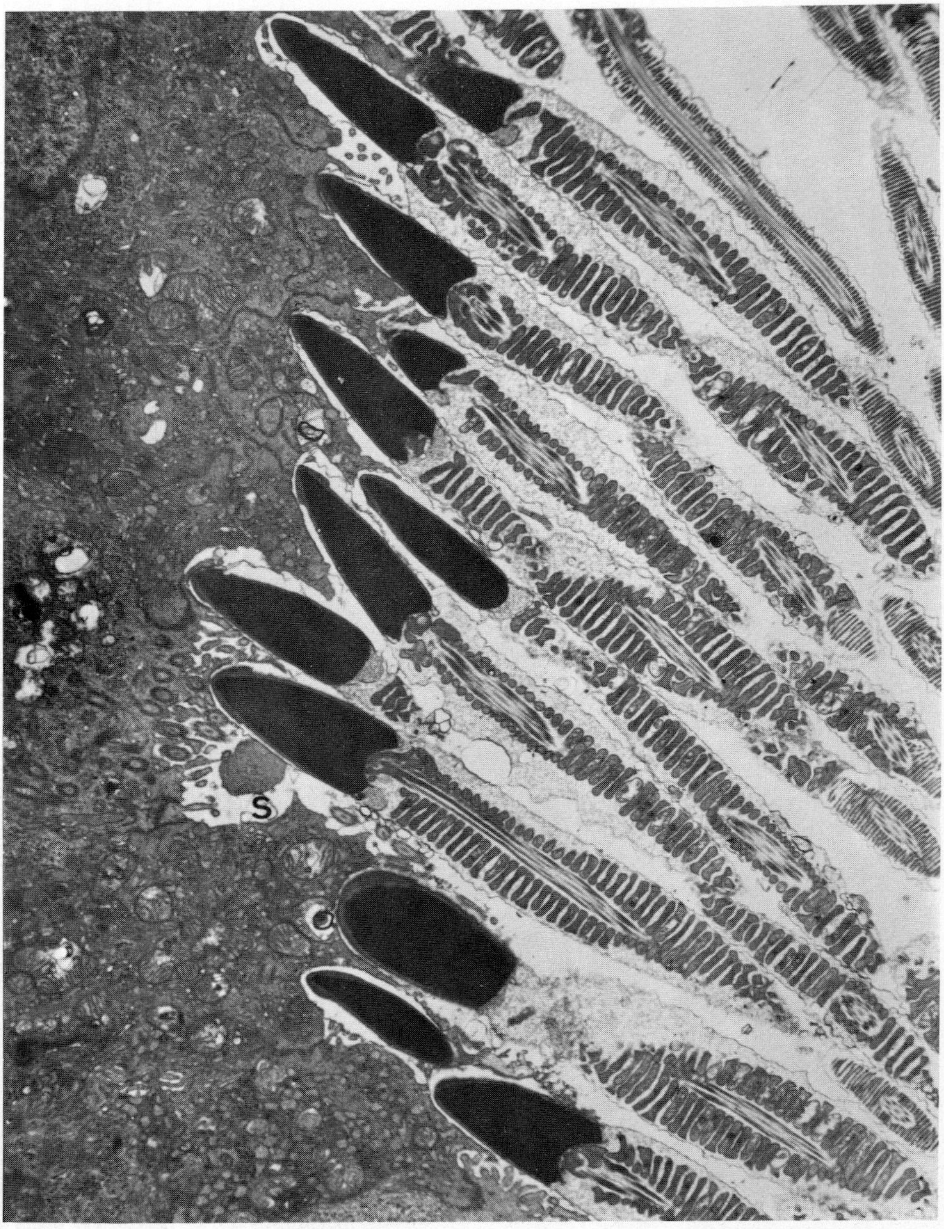

Fig. 3. Electron micrograph showing parallel orientation of sperm towards epithelial cells in the junctura region of the oviduct of *Tylonycteris pachypus.* The sperm heads lie in identations of the epithelial plasma membrane. Secretory material (s) is applied to the plasma membrane of a sperm head. × 7,500.

plasma membrane of the epithelial cell (fig. 4, 5). A relatively uniform gap of about 15 nm separates these membranes and is filled with an amorphous substance. The plasma membrane of the sperm is separated from the outer acrosomal membrane, but the intervening space is electron transparent and irregular, indicating that it may be a fixation artefact. It is, however, evident that sperm heads are brought into very close juxtaposition to the cytoplasm and organelles of the epithelial cells (fig. 6).

C. The Oviducal Epithelium

The oviducal epithelium is one cell thick, consisting of ciliated cells interspersed with non-ciliated cells. There are regional variations in the electron opacity of the different cell types, for in some levels of the oviduct, ciliated cells are darker while in others they are lighter. Apart from the presence of cilia, both cells have similar structures (fig. 2). They are about 15 μm high, and rest on a highly irregular basal lamina. The Golgi apparatus is prominent and both rough endoplasmic reticulum and free ribosomes are abundant. Osmiophilic bodies, similar in size to mitochondria are more commonly found in non-ciliated cells, and may be secretory granules. Small vacuoles are occasionally seen immediately below the apical plasma membranes of the non-ciliated cells. The lateral cell membrane of both ciliated and non-ciliated cells appears straight for most of its height with some folding at its base. The apical parts of adjacent cells are joined with junctional complexes.

Very little PAS-positive material was located in the oviducts examined; a thin layer was seen at the luminal edge of the epithelial cells, and granular material in the lumen of others. Best's carmine stain for glycogen gave negative results. Sudanophilic lipids were concentrated in the apical border of the oviducal epithelial cells along the length of the oviduct (14) and pyronin-positive ribonuclease-labile material was distributed throughout the epithelia, confirming the presence of RNA. The dinitrofluorobenzene reaction for protein showed a diffuse reaction throughout the cells, but no subcellular concentration.

IV. Discussion

Orientation of spermatozoa with their heads towards the epithelial lining of their storage organs has previously been described in hibernating bats (1, 16). *Racey and Potts* (13) drew attention to contact between the microvilli of the uterine epithelium and the heads of adjacent spermatozoa in the European pipistrelle, and *Napolitano et al.* (10) and *Wimsatt et al.* (18) reported that some

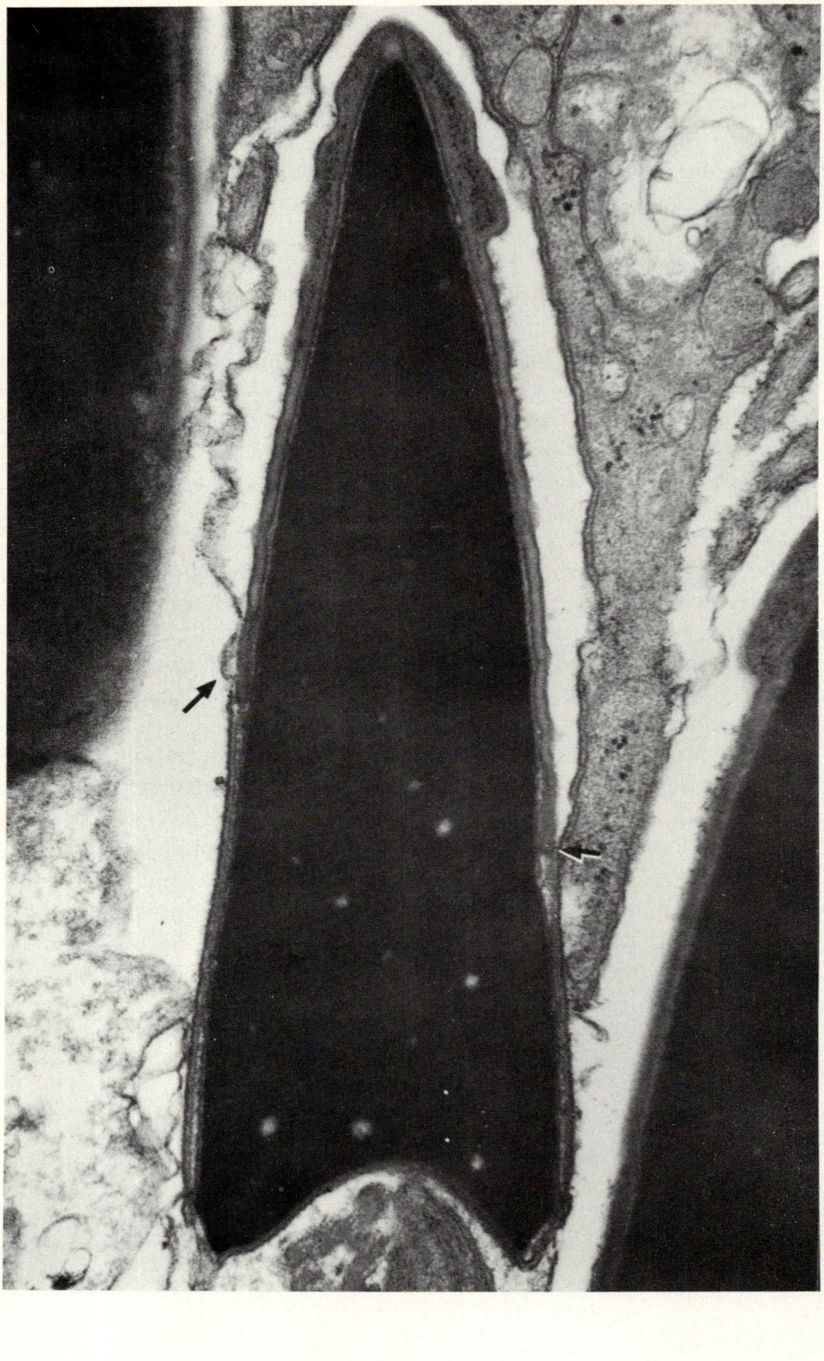

4

sperm were embedded in uterine epithelial cells in the North American vespertilionid *Myotis lucifugus.*

The present study has revealed that the relatively few sperm stored in the oviduct of *Tylonycteris* are all in close contact with the epithelium. Individual spermatozoa lie in indentations in the epithelial plasma membrane, so that intimate contact is established between it and the plasma membrane of the sperm. Although the separation that has occurred between the plasma membrane and outer acrosomal membrane of the spermatozoa in the present work may be due to high osmolarity of the fixative (6), the presence of an amorphous substance and the constant distance of separation between the two opposed plasma membranes indicates that such apposition is authentic and not a fixation artefact. As well as bringing the sperm of *Tylonycteris* into close proximity to potential sources of exogenous nutrients, the relationship that they establish with the epithelium may serve to anchor them and prevent their loss from the reproductive tract.

Close relationships are also established between spermatozoa and their storage organs in other vertebrates. In the domestic hen, spermatozoa are stored with their heads embedded in microvilli originating from the oviducal epithelial cells (17). Sperm heads often indent the epithelial plasma membrane of the sperm-storage receptacles in the oviduct of the garter snake *Thamnophis sirtalis* (4). Spermatozoa stored in ovarian diverticula of the guppy *Poecilia* are also found embedded in epithelial cytoplasm (5). It seems likely that the intimate relationships that are established between spermatozoa and their storage organs in these unrelated species are associated with the prolonged survival of spermatozoa (11). In all cases, sperm are found adjacent to cell organelles or secretion products.

As far as the oviducal epithelium of *Tylonycteris* is concerned, the abundant endoplasmic reticulum, free ribosomes and osmiophilic vesicles seen in the electron microscope, correlate respectively with the pyronin-positive material and the sudanophilic lipids revealed by histochemistry. Together with the active Golgi apparatus, these features suggest active secretion by the epithelial cells.

In bats of temperate latitudes, the prolonged survival of stored spermatozoa has been attributed to the low body temperatures which prevail during hibernation (3). Evidence has, however, subsequently accrued to show that these bats frequently arouse from torpor and become active during the period of sperm storage (12, 15). The minimum ambient temperature experienced by the sympatric populations of *Tylonycteris* examined in the present work invariably

---

Fig. 4. Sperm head situated in indentation in epithelial cell. The plasma membrane is closely applied to the post-acrosomal region of the sperm head. It separates from the acrosome (arrows) and becomes closely applied to the plasma membrane of that part of the epithelial cell which extends up one side of the sperm head. × 60,000.

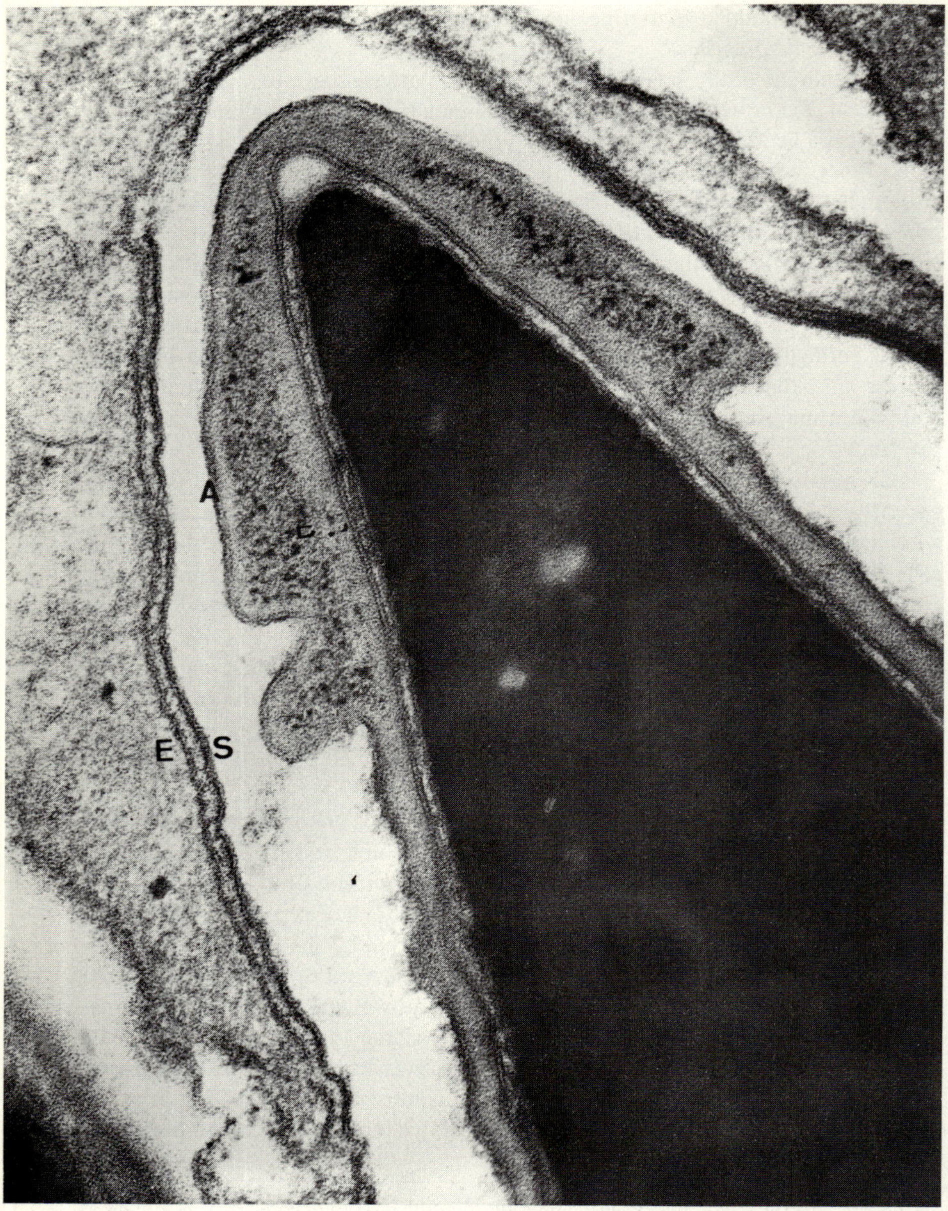

Fig. 5. Electron micrograph showing the relationship between sperm and cell membranes. E = Plasma membrane of epithelial cell; S = plasma membrane of sperm head; A = outer acrosomal membrane. × 150,000.

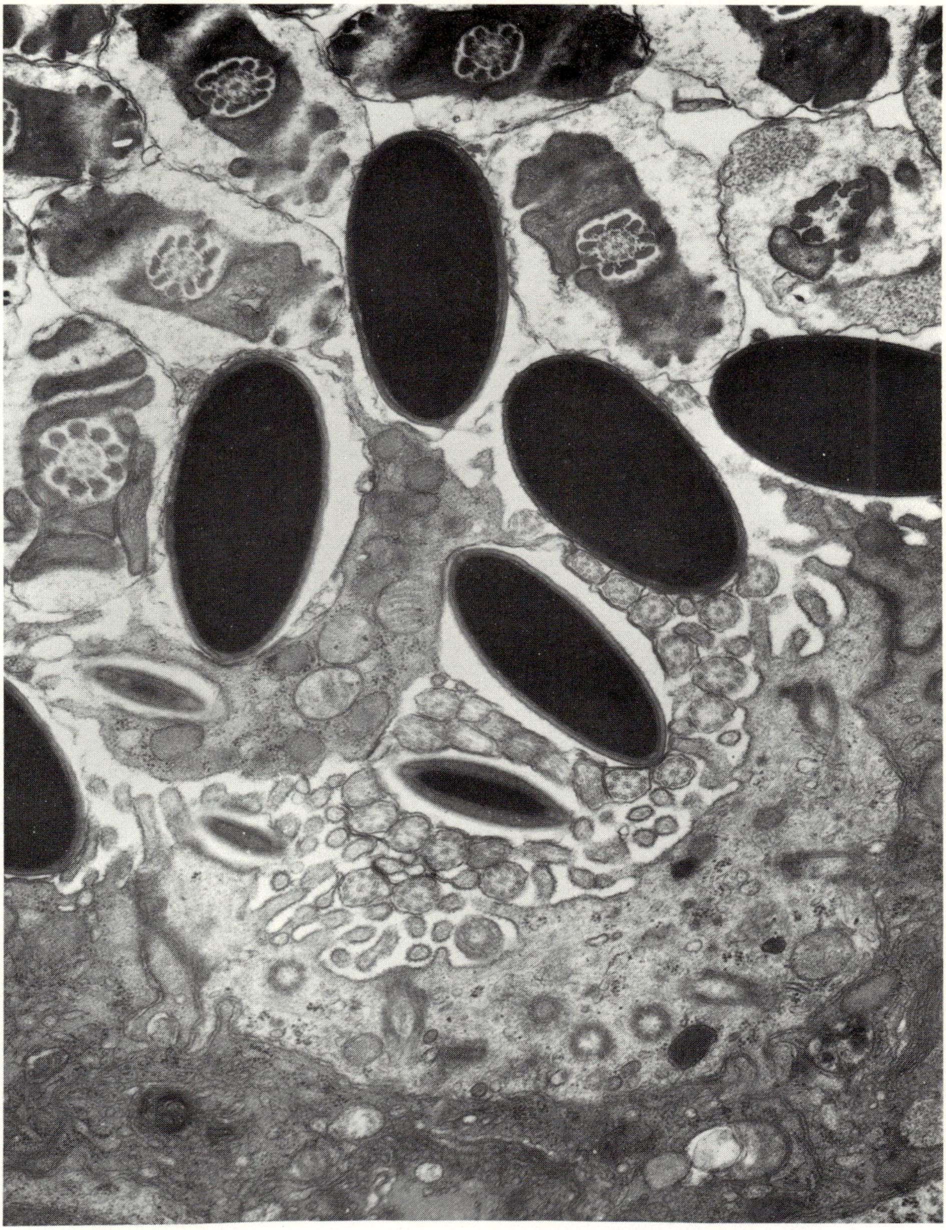

Fig. 6. Electron micrograph showing close juxtaposition between sperm heads and epithelial cell organelles. × 30,000.

exceeds 20 °C (7). It has, however, recently been established that both *Tylonycteris* species are heterothermic and, in the laboratory, inseminated females became hypothermic at an ambient temperature of 23–26 °C (8). Sperm storage and heterothermy thus appear to be associated in vespertilionid bats.

The selective advantage of sperm storage in tropical bats is not fully elucidated. For the equatorial populations of *Tylonycteris robustula* and *T. pachypus* from which specimens were taken, *Medway* (7) has suggested that it may provide one stage in an otherwise endogenous cycle when an extero-receptive factor can act as the proximate stimulus to ovulation and thus maintain the impressive synchrony of the annual reproductive cycle. The same advantage may exist in bats of temperate zones. In any case, whatever the ultimate advantage of sperm storage, it is clear that the mechanisms by which it is achieved vary considerably among the family Vespertilionidae.

References

1 *Courrier, R.:* Sur le rôle physiologique des sécrétions utérine et tubaire chez la chauve-souris hibernante. C.R. Soc. Biol. *84:* 572–574 (1921).

2 *Gopalakrishna, A. and Madhavan, A.:* Survival of spermatozoa in the female genital tract of the Indian vespertilionid bat, *Pipistrellus ceylonicus chrysothrix* (Wroughton). Proc. Indian Acad. Sci. Ser. B *73:* 43–49 (1971).

3 *Hartman, C.G.:* On the survival of spermatozoa in the female genital tract of the bat. Quart. Rev. Biol. *8:* 185–193 (1933).

4 *Hoffman, L.H. and Wimsatt, W.A.:* Histochemical and electron microscopic observations on the sperm receptacles in the garter snake oviduct. Amer. J. Anat. *134:* 71–96.

5 *Jalabert, B. et Billard, R.:* Etude ultrastructurale du site de conservation des spermatozoïdes dans l'ovaire de *Poecilia reticulata* (Poisson Téléostéen). Ann. Biol. anim. *9:* 273–280 (1969).

6 *Jones, R.C.:* The plasma membrane of ram, boar and bull spermatozoa. J. Reprod. Fertil. *33:* 179–183 (1973).

7 *Medway, Lord:* Reproductive cycles of the flat-headed bats *Tylonycteris pachypus* and *T. robustula* (Chiroptera, Vespertilioninae) in a humid equatorial environment. Zool. J. Linn. Soc. Lond. *51:* 33–61 (1972).

8 *Medway, Lord:* Hypothermia in flat-headed bats (Chiroptera, Vespertilionidae: genus *Tylonycteris*). Mammalia *37:* 646–653 (1973).

9 *Medway, Lord and Marshall, A.G.:* Roost-site selection among flat-headed bats (*Tylonycteris* spp.). J. Zool., Lond. *161:* 237–245 (1970).

10 *Napolitano, L.M.; Krutzsch, P.H., and Wimsatt, W.A.:* Electron microscopy of the sperm-uterine relationship in a hibernating bat. Anat. Rec. *145:* 365 (1963).

11 *Racey, P.A.:* The prolonged survival of spermatozoa in bats; in *Duckett and Racey* The biology of the male gamete. Biol. J. Linn. Soc. Lond. *6:* Suppl. 1.

12 *Racey, P.A.:* The time of onset of hibernation in pipistrelle bats. *Pipistrellus pipistrellus.* J. Zool., Lond. *171:* 465–467 (1973).

13 *Racey, P.A. and Potts, D.M.:* The relationship between stored spermatozoa and the uterine epithelium in the pipistrelle bat *(Pipistrellus pipistrellus).* J. Reprod. Fertil. *22:* 57–63 (1970).

14 *Racey, P.A.; Suzuki, F., and Medway, Lord:* The relationship between stored spermatozoa and the oviducal epithelium in bats of the genus *Tylonycteris;* in *Thibault and Hafez* Transport et survie et pouvoir fécondant des spermatozoïdes chez les vertébrés (INSERM, Paris 1973).
15 *Ransome, R.:* The effect of ambient temperature on the arousal frequency of the hibernating greater horseshoe bat *(Rhinolophus ferrumequinum),* in relation to site selection and the hibernation site. J. Zool., Lond. *164:* 353–371 (1971).
16 *Redenz, E.:* Das Verhalten der Säugertierspermatozoen zwischen Begattung und Befruchtung. Z. Zellforsch. mikrosk. Anat. *9:* 734–749 (1929).
17 *Van Krey, H.P.; Ogasawara, F.X., and Pangborn, J.:* Light and electron microscopic studies of possible sperm gland emptying mechanisms. Poult. Sci. *46:* 69–78 (1967).
18 *Wimsatt, W.A.; Krutzsch, P.H., and Napolitano, L.:* Studies on sperm survival mechanisms in the female reproductive tract of hibernating bats. I. Cytology and ultrastructure of intra-uterine spermatozoa in *Myotis lucifugus.* Amer. J. Anat. *119:* 25–60 (1966).

Authors' addresses: Dr. *P.A. Racey,* Department of Zoology, University of Aberdeen, *Aberdeen, AB9 2TN* (Scotland); Lord *Medway,* Great Glemham House, *Saxmundham, Suffolk, IP17 ILP* (England); Dr. *Fumie Suzuki,* Department of Anatomy, School of Medicine, University of Chiba, *Chiba 280* (Japan).

The Biology of Spermatozoa
INSERM Int. Symp., Nouzilly 1973, pp. 134–144 (Karger, Basel 1975)

Prolonged Spermatozoan Survival in the Female Hare Uterus: Explanation of Superfetation

L. Martinet and F. Raynaud
(with the technical assistance of *R. Monnerie and V. François*)
Station de Physiologie animale, INRA and Université de Paris VI, Jouy-en-Josas

Introduction

In the female hare *(Lepus europaeus),* gestation lasts 41–42 days, but the interval between two successive parturitions in a female left permanently with a male is often less than 41 days (10). The few cases reported by *Hediger* and his collaborators have been confirmed in other observations (12). In 60% of the cases, the interval between two successive births was less than the length of gestation, and varied between 35 and 40 days, the average being 38 days (12).

This observation indicated possible superfetation: the result of fertilization and development of eggs in an already pregnant female, or the simultaneous development of two different generations of eggs. Such superfetation has been described since antiquity, and has been confirmed by observation of the sexual chase and mating followed by fertilization several days before parturition (13).

For many authors, this phenomenon could only be explained by unilateral gestation, rather frequently occurring in this species in which the uterine horns function alternatively (8, 11, 18).

The case described by *Bloch et al.* (1) confirmed this hypothesis. In fact, the first gestation having been unilateral, it could be supposed that spermatozoa had ascended in the empty uterine horn and passed by transperitoneal pathway to fertilize the oocytes on the opposite side, as had been described in other species (6, 16).

However, discovery of superfetation when the first gestation was bilateral, again raised the question of the mechanism of this phenomenon (12).

Experiment and Results

These mechanisms may be involved: either passage of spermatozoa in a uterus containing almost full-term fetuses, or storage of spermatozoa in the genital tract beyond the implanted fetuses, after the previous mating with prolonged survival of spermatozoa in the female genital tract. This was maintained and checked by substituting a vasectomized male for a fertile male.

Prolonged Survival of Spermatozoa in the Genital Tract of the Female Hare

A female hare giving birth on March 19 was isolated from the normal male on March 23. A vasectomized male was introduced in her pen on April 20, and on April 22 mating was observed.

The female gave birth to three young hares on April 25. Laparotomy performed a few hours later showed two freshly ruptured follicles on the right ovary and one on the left; two cleaved eggs were recovered by flushing the right oviduct (fig. 1 a). Some spermatozoa have thus kept their fertilizing ability for at least 30 days (March 23 to April 22).

An identical case showed sperm survival of at least 10 days (fig. 1b), since the normal male was taken away at the beginning of tha time. Survival in this case was certainly longer and should have corresponded to the gestation period.

Site of Spermatozoan Storage

Shorter or longer periods of prolonged survival of spermatozoa in the female genital tract of mammals occurs in the cervix, uterine glands, utero-tubal junction, isthmus or ampulla, depending on the species (19).

In the superfetation process, spermatozoa should be stored between fetal implantation sites and the site of fertilization. We systematically studied the oviduct and uterus at 48 h, 10 and 17 days of pregnancy in 3 hares isolated from the male for 48 h, 6 and 14 days, respectively.

Anatomy of the Uterus

The uterus is bicornate. The two horns open individually in the vagina.

In the nonpregnant female, the endometrium is seen as four large longitudinal folds; at the bottom of the troughs there are fewer glands than at their apex.

As the tube opens into the uterine cavity the isthmic mucosa terminates in long processes, which project into the uterus and are free of glands.

At 48 h of pregnancy, the following can be seen in the endometrium

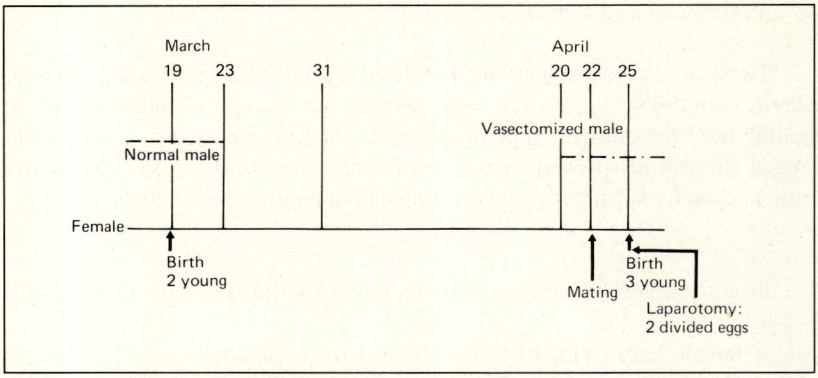

Fig. 1 a. Observation of prolonged spermatozoan survival.

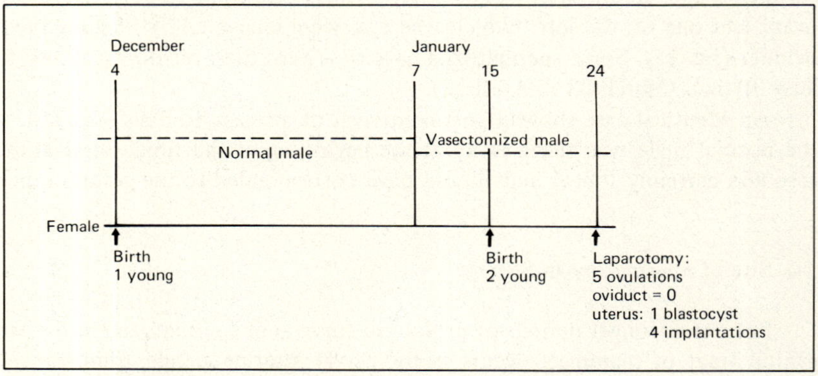

Fig. 1 b. Observation of prolonged spermatozoan survival.

(fig. 2a): (1) A monolayered columnar or pseudostratified epithelium; its cells are tall with elongated nuclei situated in a basal position. (2) A stroma divided into three areas: a relatively dense area occupied by the neck of the uterine glands; a loose edematous area including the distal or 'blind' end of the glands; and, against the myometrium, a slightly denser area free of glands. (3) Straight or slightly coiled uterine glands, perpendicular to the longitudinal axis of the endometrium, 250–350 μm long, and delimited by a monostratified epithelium (fig. 2b).

At 10 and 17 days of pregnancy, the three areas of the stroma are no longer recognizable. They are completely invaded by uterine glands, which are extremely coiled and branched. These glands reach the myometrium (fig. 2c).

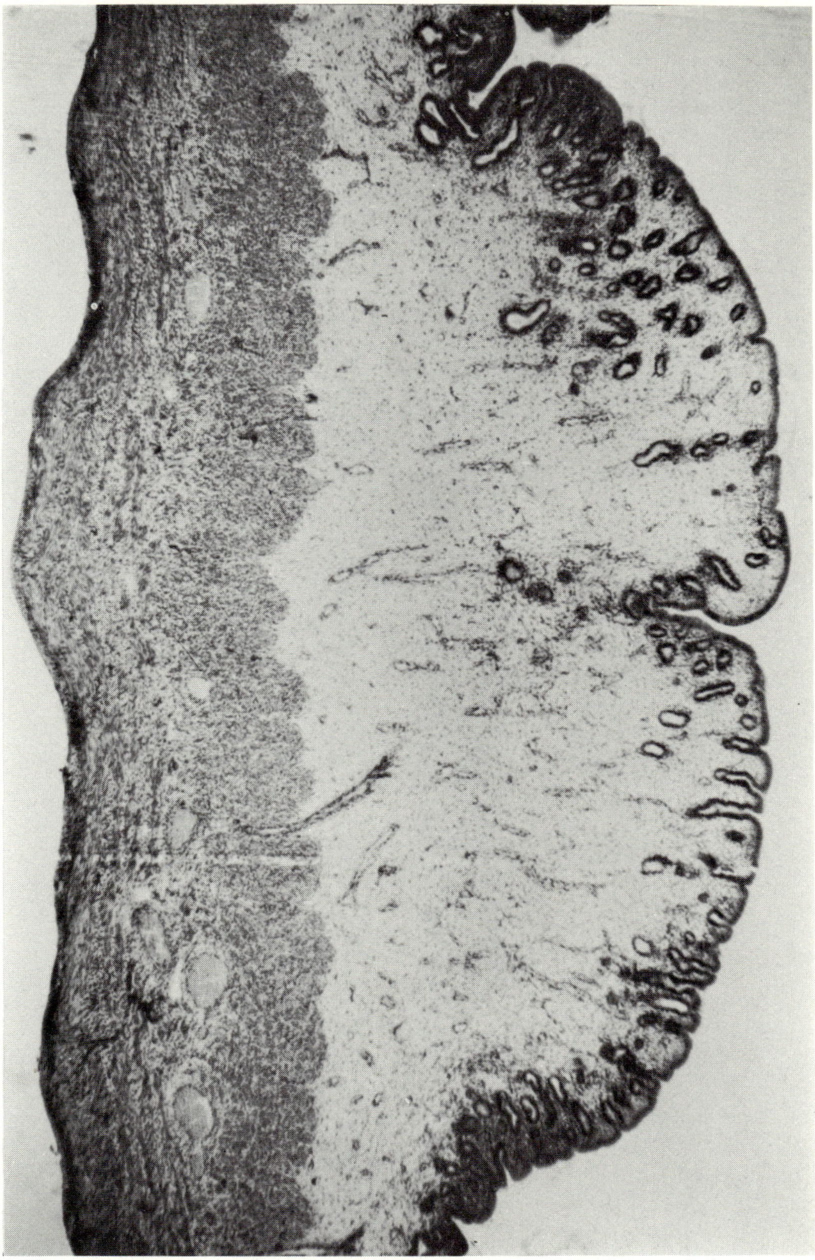

Fig. 2a. Longitudinal section of uterus. Gestation 48 h.

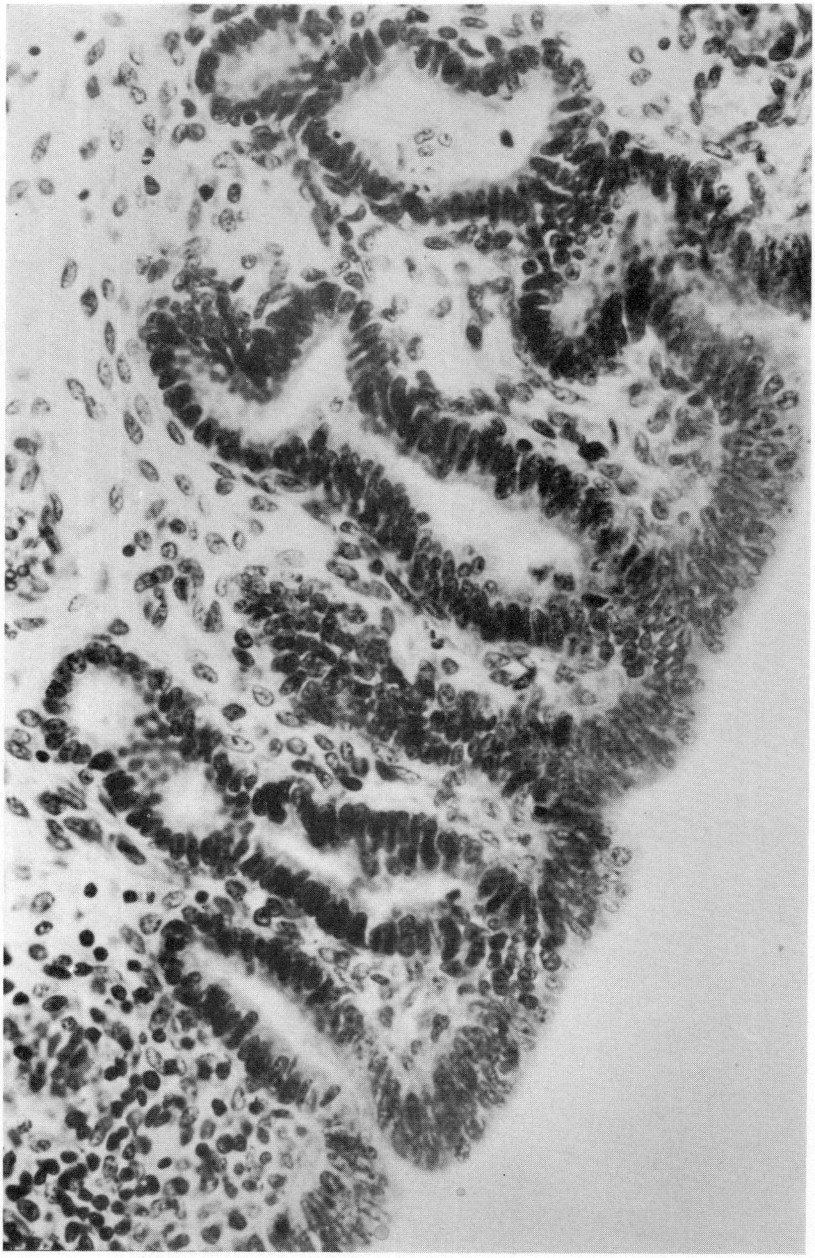

Fig. 2b. Straight or slightly coiled uterine glands. Gestation 48 h.

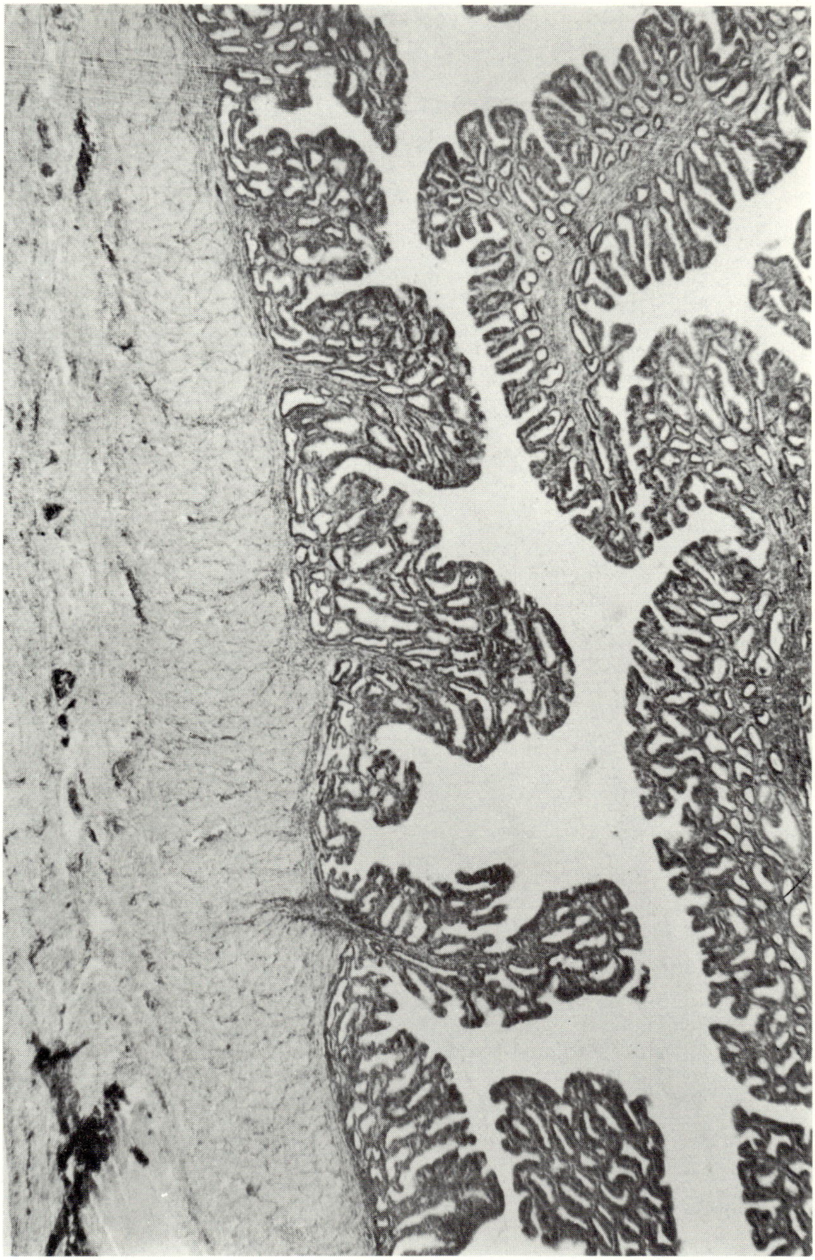

Fig. 2c. Longitudinal section of uterus. Gestation 10 days.

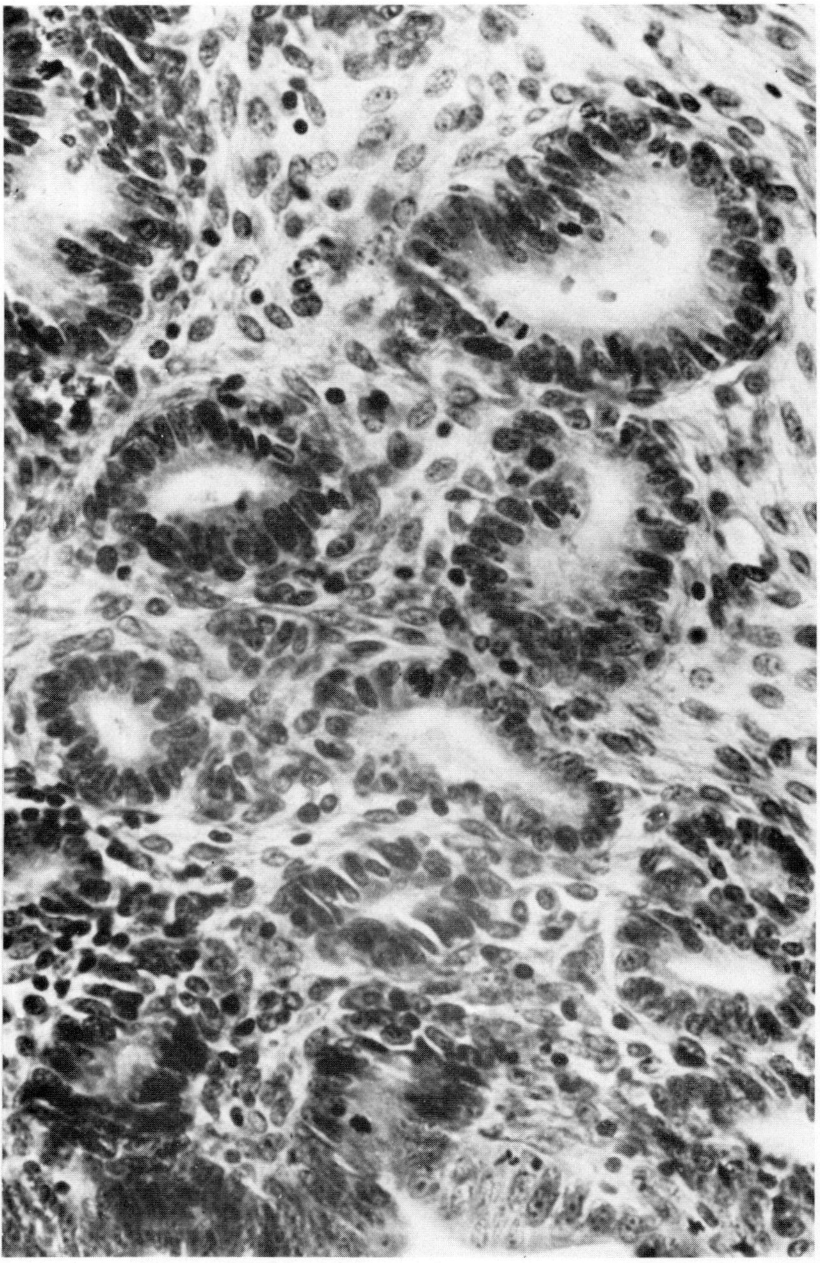

Fig. 2d. Spermatozoa in the bottom of uterine gland having 5 spirals.

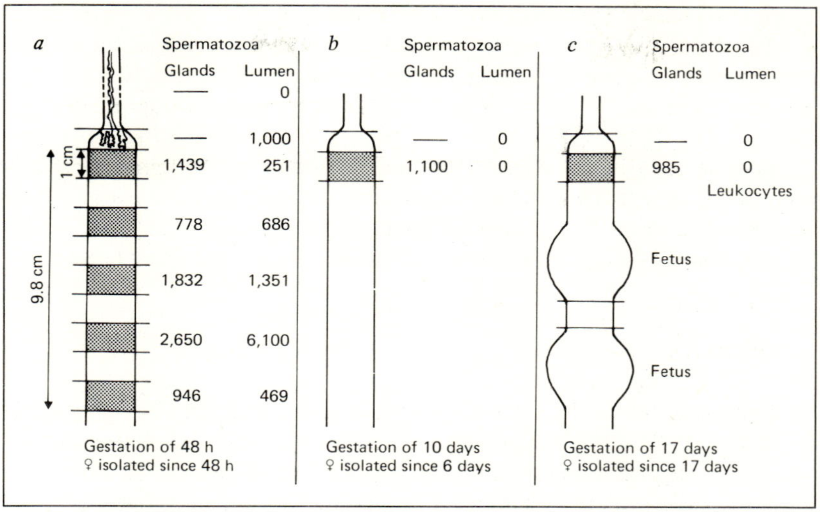

Fig. 3a, b, c: Spermatozoan count in the female-hare uterus at 3 different stages of pregnancy.

Distribution of Spermatozoa in the Uterus and Oviduct 48 h after Mating

A female hare was inseminated by vaginal route with 9.5×10^5 spermatozoa, and the right uterus and oviduct were removed 48 h later.

Spermatozoan counts were made on longitudinal sections of the whole oviduct and on five portions 1 cm long, distributed regularly over 9.8 cm of the uterus (fig. 3a).

No spermatozoa were found in the oviduct, even though their transport was normal. This was confirmed by the presence of a cleaved egg in the oviduct, and the birth of a young hare 39 days later from the left portion of the tract that had been left in place.

44 % of the 17,500 spermatozoa counted in the uterus were in the glands, the remainder were found in the uterine lumen. A large number (1,000) was accumulated on the epithelial folds of isthmic projections. Spermatozoan distribution along the uterus did not show any privileged zone of penetration into the glands (fig. 3a).

The total number of spermatozoa in the uterus, estimated by the extrapolation of their number counted over 6 cm, represented 3 % of the number inseminated.

In the glands, spermatozoa had always penetrated far into the distal end (fig. 2d). However, they were never found in clusters, but in twos and threes in the gland lumen. The short or noncoiled glands did not contain spermatozoa.

Spermatozoan Survival at the Utero-Tubal Junction

The number of spermatozoa present in the uterine glands at the utero-tubal junction was relatively constant during pregnancy: 1,439, 1,100 and 985 on segments 1 cm long at 48 h, 10 and 17 days of pregnancy, respectively (fig. 3a, b, c).

Spermatozoa in the glands were always found in the lumen of the gland, rarely lying against the glandular epithelium. No polymorphonuclear leukocytes were observed in the lumen or in the subepithelial stroma

At 10 and 17 days of pregnancy the uterine lumen was free of sperm.

Discussion

Superfetation in the hare is not exceptional, since 18 % of pregnancies in the field (8) and 60 % in captivity (12) occur as a result of prepartum mating. Moreover, this phenomenon does not represent an adaptation for survival in the species because ovulation is induced by mating and a male is necessary, even if living sperm are present in the uterus.

This study shows that superfetation can be explained by the survival and preservation of spermatozoan fertilizing ability for at least one month in the uterus. Following is a chronology of this phenomenon: a first mating allows fertilization of a first series of eggs and the storage of spermatozoa in the uterine glands; a second mating a few days before parturition induces a further ovulation. The eggs are thus fertilized by the spermatozoa inseminated during the first mating (fig. 4). However, what happens at the third gestation? Two hypotheses may be considered: the uterus still contains a sufficient number of sperm, which have maintained their fertilizing ability; or a mating is necessary after the second parturition to restock the uterine glands with spermatozoa. The second alternative is possible because postpartum heat is observed if it has not appeared before

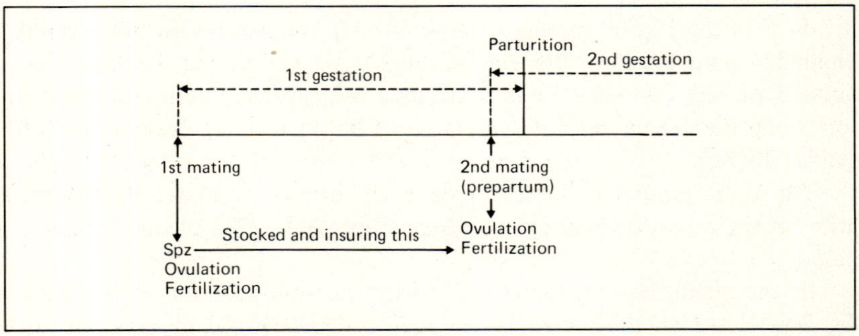

Fig. 4. Chronology of events in superfetation.

parturition. Contrary to the camelides (17) or the common shrew (2), the spermatozoa are not stored in the oviduct. Their disappearance seems even faster than in the rabbit, where 48 h after insemination 100–300 spermatozoa still remain (7).

On the other hand, the rate of disappearance in the entire tract is identical in the two species: 1 % of the spermatozoa are still present 48 h after artificial insemination.

Thus, in the hare, as in several species such as dog (5), hedgehog (4), bat (3, 15), the uterine glands are the site of storage and survival of spermatozoa.

However, in the hare we have never found spermatozoa facing toward the uterine epithelium, as *Racey and Potts* (15) noticed in the bat.

There does not seem to be a privileged site in the uterus with any specific power of attraction, as spermatozoa are evenly distributed all along it. However, if there is a higher gland density in the distal part of the uterus, as shown by measurements on the sow (14), the glands of the utero-tubal portion would then attract spermatozoa more strongly than the others. In any case, it would appear that spermatozoa are in glands which demonstrate some coiling 48 h after mating; short straight glands are always empty.

We do not know yet if survival is better in the glands of the cranial part of the uterus than in the middle segment (e.g. between fetus implantation sites), or in the lower part. As yet, spermatozoa have not been counted at these levels after long preservation times.

Finally, we still do not know about the mechanism by which spermatozoa migrate from the uterine glands towards the oviduct at the second mating or prepartum ovulation. The release of luteinizing hormone does not seem to be involved; none of the ovulated eggs recovered from seven hares, injected with HCG 2 or 3 days before parturition, were fertilized. Oxytocin, which might be released at mating time, does not seem to have taken part either. Four hares, which received both 50 IU of HCG and 0.5 IU of oxytocin, ovulated without fertilization of their eggs.

The idea of the contraction of myoepithelial cells surrounding the glands, which has been postulated in the hen (9) does not seem applicable to the hare. Moreover, myoepithelial cells have never been found around the uterine glands, even in the hen.

In conclusion, superfetation occurs very frequently in the hare because 3 conditions are fulfilled at the same time in this species: estrus behavior, prepartum ovulation, and prolonged spermatozoan survival in the uterine glands.

References

1 *Bloch, S.; Hediger, H.; Lloyd, H.G.; Muller, G. und Strauss, F.:* Beobachtungen zur Superfetation beim Feldhasen. Z. Jagdwiss. *13:* 49–51 (1967).

2 *Brambell, F.W.G.:* Reproduction in the common shrew (*Sorex araneus* L.). I. The oestrous cycle of the female. Phil. Trans. roy. Soc. B *225:* 1–49 (1935).

3 *Courrier, R.:* Sur le rôle physiologique des sécrétions utérine et tubaire chez la chauve-souris hibernante. C.R. Soc. Biol. *84:* 572 (1921).

4 *Deanesley, R.:* The reproductive processes of certain mammals. VI. The reproductive cycle of the female hedgehog. Phil. Trans. roy. Soc. B *223:* 239–276 (1934).

5 *Doak, R.L.; Hall, A., and Dale, H.E.:* Longevity of spermatozoa in the reproductive tract of the bitch. J. Reprod. Fertil. *13:* 51–58 (1967).

6 *Edgar, D.G. and Asdell, S.A.:* Spermatozoa in the female genital tract. J. Endocrin. *21:* 321–326 (1960).

7 *El Banna, A.A. and Hafez, E.S.E.:* Sperm transport and distribution in rabbit and cattle female tract. Fertil. Steril. *21:* 534–540 (1970).

8 *Flux, E.C.:* Reproduction and body weights of the hare, *Lepus europaeus* in New Zealand. New Zeald J. Sci. *10:* 357–401 (1967).

9 *Gilbert, A.B.; Reynolds, M.E., and Lorenz, F.W.:* Distribution of spermatozoa in the oviduct and fertility in domestic birds. VII. Innervation and vascular supply of the utero-vaginal sperm host glands of the domestic hen. J. Reprod. Fertil. *17:* 305–310 (1968).

10 *Hediger, H.:* Die Zucht des Feldhasen *(Lepus europaeus)* in Gefangenschaft. Physiol. Comp. Oecol. *1:* 46–62 (1948).

11 *Knauss, H. von:* Zur Frage der Superfetation beim Feldhasen. Z. Jagdwiss. *13:* 52–53 (1967).

12 *Martinet, L.; Leguois, J.J. et Moret, B.:* Quelques observations sur la reproduction du lièvre européen *(Lepus europaeus)* en captivité. Ann. Biol. anim. Biochem. Biophys. *10:* 195–202 (1970).

13 *Martinet, L. et Raynaud, F.:* Mécanisme possible de la superfœtation chez la hase. C.R. Acad. Sci. *274:* 2683–2686 (1972).

14 *Pelletier, M.C.:* Personal commun. (1967).

15 *Racey, P.A. and Potts, D.M.:* Relationship between stored spermatozoa and the uterine epithelium in the pipistrelle bat *(Pipistrellus pipistrellus)*. J. Reprod. Fertil. *22:* 57–63 (1970).

16 *Rowlands, I.W.:* Insemination by intraperitoneal injection. Proc. Soc. Stud. Fertil. *10:* 150–157 (1958).

17 *Stekleniov, E.P.:* Des particularités anatomo-morphologiques de la structure et des fonctions physiologiques des trompes de Fallope chez les camélidés (en russe). 6e Congr. int. Reprod. Anim. Insém. Artif., Paris 1968, vol. 1, pp. 71–74.

18 *Stieve, H. von:* Zur Fortpflanzungsbiologie der europäischen Feldhasen *(Lepus europaeus)*. Zool. Anz. *148:* 101–114 (1952).

19 *Thibault, C. et Levasseur, M.C.:* Conservation et survie prolongée des spermatozoïdes dans les voies génitales femelles des vertébrés. Ann. Biol. anim. Biochem. Biophys. *13:* 267–284 (1973).

Author's address: *Lise Martinet and Françoise Raynaud,* Institut National de la Recherche Agronomique, Station Centrale de Physiologie animale, *F–78350 Jouy-en-Josas* (France)

The Biology of Spermatozoa
INSERM Int. Symp., Nouzilly 1973, pp. 145–155 (Karger, Basel 1975)

Transport, Migration and Survival of Spermatozoa in the Female Genital Tract: Species with Intra-Uterine Deposition of Semen

R.H.F. Hunter

Department of Agricultural Zoology, School of Agriculture, University of Edinburgh, Edinburgh

Introduction

This chapter summarizes important aspects of the processes of sperm transport, migration and storage in the female tract of four species of mammal in which much of the ejaculate enters the uterus during coitus. A more detailed treatment and list of references concerning the fate of spermatozoa in the female reproductive tract is given in the recent volume by *Thibault and Hafez* (27). Before describing similarities and differences on these topics among the four species, rat, dog, horse and pig, a number of basic facts relevant to the arrival of a population of competent spermatozoa at the site of fertilization must be presented.

First, it is necessary to distinguish between *sperm transport,* which implies a passive movement of cells in the genital tract, and *sperm migration* which clearly attributes great importance to the intrinsic motility of the cell. Although early investigators were inclined to view sperm ascent along the female tract as an outcome of the sperm cells' own movements, it is now generally accepted that the role of sperm motility becomes paramount only in certain regions of the tract (e.g. cervix and utero-tubal junction) and during penetration of the egg membranes. The rapidity with which spermatozoa appear in the upper portions of the Fallopian tubes after mating or artificial insemination has emphasized that sperm motility alone could not have contributed significantly to passage over the relatively large distances involved. The journey is probably effected at all stages by contractile mechanisms in the musculature of the tract, part of the transportation process undoubtedly being assisted by central nervous mechanisms set in motion by the sexual stimulation attending coitus. Nonetheless, when the duration of oestrus is long and mating takes place early, active migration of spermatozoa from storage regions such as the lower isthmus must be of physiological significance, especially since the enhanced muscular contractions of the uterus and tubes promoted by oxytocin released at coitus would long since have ceased. As a corollary, it may be inferred that rapid post-coital transport of spermatozoa to the ampulla in this latter situation would have had little influence on fertility.

Second, in comparison with the number of spermatozoa deposited in the female tract at mating, there is a massive reduction in sperm numbers as one proceeds towards the ampulla of the Fallopian tube. This has been reasoned to reduce the chances of polyspermy by restricting the numbers of spermatozoa that achieve the site of fertilization. Third, during this marked diminution in sperm numbers, the cells are progressively removed from the seminal plasma, and their economy becomes strongly influenced by the luminal fluids of the uterus and Fallopian tubes. This situation appears to have a direct bearing on the metabolic activity of ejaculated spermatozoa, and may also be associated with development of the capacitated state.

Before proceeding to a more detailed discussion of sperm transport, the evolutionary or biological significance of intra-uterine deposition of the ejaculate deserves consideration. This site of ejaculation could be viewed as an adaptation to accommodate the copious production of seminal fluid by the male. More plausible, however, would be the argument that the intra-uterine deposition of semen in large volumes is a feature that should contribute to successful passage of spermatozoa to the site of fertilization. If this interpretation is correct, then for reasons not readily apparent, cervical and uterine transit may have proved a formidable undertaking for spermatozoa in ancestral forms of the four species discussed below. In these species, the regulatory and storage roles enjoyed by the cervix in vaginal depositors such as the rabbit, ruminants and primates appear to have been taken over in large measure by uterine glands, the utero-tubal junction and lower regions of the Fallopian tube. These remarks notwithstanding, species differences in the survival time of spermatozoa do not appear to be correlated with the site of deposition of semen at coitus, or with its mode of transport in the female tract.

Rat

As in other myomorph rodents, copulation in the rat commences with a series of intermittent thrusting movements by the male. Tonic contractions of the circular muscle of the cervix during oestrus are responsible for the retention of uterine fluids, which normally distend the cornua at this time. The cervix apparently relaxes sufficiently at the moment of ejaculation for a small mass of compacted spermatozoa to be projected through the canal into the fluid-filled uterine horns. Accordingly, passage of spermatozoa within the distended lumen to the region of the utero-tubal junction may take only a few seconds (6), being aided by the strong waves of contraction passing along the horns. Prostaglandins have been detected in rat semen (28), and their smooth-muscle stimulating activity may enhance contractions of the myometrium at this stage. A vaginal plug that abuts firmly against the cervix is formed during ejaculation, a major function apparently being to ensure that the semen is propelled into the uterus. If the seminal vesicles and coagulating glands are removed surgically, then mating of oestrous females by such males does not lead to sperm deposition in the uterus, suggesting the importance of the vaginal plug in directing the semen through the cervical canals and subsequently sealing these structures (4).

As to numbers of spermatozoa reaching different levels of the female tract, approximately 50 million are impelled into the uterus. In comparison with this uterine population, the number of spermatozoa entering the Fallopian tubes is extremely small, there being some 30,000 times more spermatozoa in the uterus than in the tubes (6). For example, 12 h after ejaculation or several hours after ovulation, approximately 500 spermatozoa are found in each tubal isthmus. The utero-tubal junction thus provides a major barrier to sperm transport, this structure seeming to control the number of spermatozoa entering the lower portion of the tube. A further significant reduction also apparently occurs along the length of the isthmus, since only 10–15 spermatozoa can be found in each ampulla 12 h after ejaculation (6). Before penetration of the egg membranes is possible, spermatozoa of this species need to undergo capacitation occupying a period of some 2–3 h (1).

The precise manner whereby spermatozoa reach the site of fertilization remains to be clarified, but their presence in the ampullae within 15–30 min of ejaculation (5) emphasizes the contribution of muscular contractions in the tubal walls, and possibly also those of the mesosalpinx. Some disagreement still exists over the occurrence and degree of sperm selection in the female tract of rats. *Leonard and Perlman* (18) reported that not only could the utero-tubal junction selectively permit passage of live as against dead spermatozoa into the tubes, but also that following surgical insemination of mixed populations of motile spermatozoa into the uterus, only rat spermatozoa gained entry to the tubal isthmus. These findings have been disputed by several authors, small numbers of non-motile rat, guinea pig and human spermatozoa being detected in the tubes after heterologous insemination into the cornua of oestrous rats. However, in the light of *Braden and Gluecksohn-Waelsch*'s work (7) showing a selection at the utero-tubal junction against mouse spermatozoa carrying certain t alleles, a specialized function of this region in the rat should not be discounted.

As is the case in many mammalian species, the duration of motility of rat spermatozoa in the female tract exceeds their fertile life. Although *Yochem* (29) observed feeble motility among spermatozoa after 17 h in the female tract, this should be contrasted with the results of *Soderwall and Blandau* (26) involving artificial insemination of 184 rats. In this study, 'conceptions' were obtained from inseminations made as long as 14 h before ovulation, but the incidence of fertile inseminations and the litter size both decreased when the interval was greater than 10 h. It can be inferred from the report of *Blandau and Odor* (6) that the principal storage region for the spermatozoa during this interval must be in the lower portion of the Fallopian tube.

Concerning the fate of spermatozoa in the female rat, phagocytosis of these cells has been observed among the uterine contents. Leucocytic infiltration of the uterine lumen and engulfment of spermatozoa increased progressively with the interval elapsing after mating, but invasion of the uterine lumen by leucocytes was thought to be primarily due to distension of the uterus with fluid (2). The relaxation of the highly contracted cervix near the end of oestrus, and the subsequent evacuation of the uterine contents early in metoestrus, ultimately disposes of the majority of spermatozoa ejaculated into the cornua (2, 6). As to those spermatozoa that have gained the Fallopian tubes, penetration of the tubal mucosa has been recorded (3), but the fate of most spermatozoa in this portion of the tract remains undecided.

Dog

Coitus in the dog is prolonged, during which time there is a more or less continuous but slow emission of seminal fluid which is thought to be deposited almost directly through the short cervical canal into the body of the uterus. The volume of the ejaculate may exceed 15 ml in the larger breeds (unpublished observations), and although enormous fluctuations in sperm density are found, 1.25 to 2.75×10^8 cells per ml represents an average figure for the whole ejaculate. Three fractions are usually distinguishable during ejaculation: first, a clear, watery fluid without spermatozoa, followed by a more viscous, white fraction rich in spermatozoa, and finally a clear, watery secretion again containing very few or no spermatozoa (15).

Distribution of these fluids throughout the uterine horns is rapid, being promoted by myometrial contractions which enable semen to reach the ovarian end of the horns shortly after coitus. Spermatozoa may thus be transported to the base of the Fallopian tubes within a minute or so of coitus and, by means of the uterine fistula technique, have been recorded at that point within 25 sec of ejaculation (13). Passage through the utero-tubal junction and along the Fallopian tubes may also be accomplished with considerable speed, for spermatozoa have been detected in the isthmus 2 min after coitus and may be distributed throughout the tubes within 20 min. However, accurate quantitative data on sperm distribution in the isthmus and ampulla are not available.

The period of storage of spermatozoa in the reproductive tract of the bitch is relatively long, and a relationship has been suggested between the survival time and the duration of oestrus, which may extend to 7 days or more. The study of *Griffiths and Amoroso* (14) inferred that spermatozoa could retain their fertilizing power in the bitch for 84 ± 12 h after breeding, but it appears from the more recent work of *Doak et al.* (11) that motile spermatozoa may be found in the uterine lumen in undiminished concentrations for 4–6 days after mating. In the same study, three of four bitches examined 215 h after mating had motile spermatozoa in the uterine lumen, and some motile cells were found as late as 268 h after deposition.

Turning to the site of storage, the uterine glands appear to play a significant role in the maintenance of sperm cells in this species. The presence of spermatozoa in these glands was noted by *Griffiths and Amoroso* (14) in a bitch autopsied 2 days after a second service, but a specific storage function for viable cells was not adduced. In the study of *Doak et al.* (11), clumps of spermatozoa were frequently found in the neck of the uterine glands, leading to the proposition that these 'glandular' spermatozoa replace those in the lumen as the latter are lost by migration and attrition. Moving to the Fallopian tubes, spermatozoa were found here and in the ovarian bursae of a series of bitches between 20 and 40 h after copulation and some spermatozoa persisted in the tubes for 168 h (11).

In bitches examined after the onset of metoestrus, very few spermatozoa were found in either the uterine lumen or the adjoining glands (11); the disappearance of spermatozoa was thought to be associated with the onset of metoestrus, not with the period of time they had been in the uterus. However, these authors presented no observations on phagocytosis of spermatozoa in the female tract, and it remains to be clarified whether leucocytic invasion of the uterus in this species is merely delayed, or whether in fact this mechanism is in abeyance.

Horse

The mare is a polyoestrous seasonal breeder, and during oestrus the reproductive tract presents characteristic oedematous features, although the cervix is usually relaxed and flaccid (10). During copulation, the glans of the penis becomes greatly engorged and abuts against the cervical os, presumably increasing the degree of dilatation in the canal. The muscular contractions at ejaculation together with the pumping action of the penis thereby transmit much of the semen through the canal into the body of the uterus. In this situation, it is difficult to appreciate the biological necessity for the development of a negative pressure in the uterus during coitus, but such a feature has been claimed, the inference being that passage of semen into the uterus is facilitated. Of greater physiological importance might be the concentration of prostaglandin E_2 in stallion semen (24.1 ng per ml – *Poyser and Hunter,* unpublished), which would promote contractile activity in the myometrium and other smooth-muscle compartments of the female tract.

The semen of the stallion is particularly voluminous, measurements commonly ranging from 18 to 170 ml. The volume is influenced by the breed of stallion, and the composition of the semen is also subject to marked seasonal fluctuations. As is the case for the other large species discussed in this chapter, various fractions can usually be distinguished in the semen: initially, a watery, pre-sperm secretion, followed by a milky, sperm-rich fraction, and terminating with a highly viscous, gelatinous material (20). The considerable volume of fluid deposited in the female tract reduces one of the problems of sperm transport for, shortly after mating, the uterine horns of the mare including the regions adjoining the Fallopian tubes are filled with the products of ejaculation (20). Variation is also found in the density of spermatozoa in the semen, values ranging from 0.3 to 8×10^8 cells per ml. A single ejaculate may contain from 4 to 13×10^9 spermatozoa.

Among the larger species of domestic livestock, there is evidence that spermatozoa of the stallion exhibit one of the longest periods of survival in the female tract. The viability of spermatozoa in the mare has been studied by several authors, the approach being dependent upon the extended period of oestrus, with ovulation occurring towards the end of this time (10). Although few observations have been found in the literature pertaining to motility of spermatozoa after periods of storage in the female tract, the evidence for prolonged maintenance of fertilizing capacity is overwhelming. *Day* (10) recorded pregnancies in mares inseminated 5 days before ovulation and, in one instance, conception followed a sperm survival interval of 6 days. These figures received support from the subsequent study of *Burkhardt* (9), but although the latter author makes the statement that ... 'in services given early in oestrus, the sperms are quite capable of reaching the fimbriae of the Fallopian tube and maintaining fertilizing capacity for 90 h or more ...', a specific site of sperm storage was not indicated.

Concerning the fate of the seminal plasma and spermatozoa, only relatively small volumes of fluid (3–5 ml) can be recovered from the uterine horns when 30–50 min have elapsed since mating, and the concentration of spermatozoa suspended in this fluid may have been reduced significantly (20). Moreover, although there is chemical evidence suggesting that very small amounts of seminal plasma may negotiate the utero-tubal junction and enter the Fallopian tubes within 50 min of mating (20), the fate of most ejaculated fluid remains uncertain; retention in the endometrial crypts and/or leakage through the cervical canal seem reasonable possibilities. Whether the prolonged periods of sperm storage are associated with specific endometrial glands, or with persistence in the lower portions of the Fallopian tube, also remains to be studied systematically.

Pig

Because a more comprehensive review of physiological aspects of sperm transport, migration and storage in the domestic pig has been presented elsewhere (16, 17), the paragraphs that follow contain only an extremely brief summary from which most of the references have been omitted.

Coitus in this species is a protracted affair, on occasions lasting for 7–8 min or more. During this period, semen is propelled by muscular contractions of the male tract almost directly into the pars indivisa of the bicornuate uterus, although passing contact is made with the proximal region of the cervical canal. The considerable volume of the ejaculate, frequently amounting to nearly half a litre, is emitted in fairly distinct fractions, commencing with a watery pre-sperm secretion, followed by a sperm-rich fraction, and terminating with a post-sperm fraction. Together with the gelatinous portion of the ejaculate which serves to occlude much of the cervical canal, this sequence of seminal fractions may then be repeated in a second wave. The fluctuation in sperm density during the prolonged ejaculation has been considered significant for efficient transport of spermatozoa to the Fallopian tubes (12); nonetheless, on account of the volume of the ejaculate, the utero-tubal junction may be bathed by semen before the end of mating. The total number of spermatozoa produced by the boar in a single ejaculate may be as high as 7.83×10^{10} cells (25), with a concentration varying between 0.1 and 10×10^8 cells per ml (22).

The rapid passage of semen to the utero-tubal junction clearly explains the presence of spermatozoa in the upper reaches of the Fallopian tubes within an hour or so of mating. The shortest transit to be documented for boar spermatozoa to the upper half of the tubes is a period of 15 min after mating (8), and sperm penetration of the zona pellucida has been noted within 2 h of intracervical insemination. But as mating under natural conditions would precede ovulation by a number of hours, the biological importance of such rapid transport of spermatozoa to the site of fertilization is questionable.

Numerous studies in gilts and sows indicate that only a very small proportion of the massive number of spermatozoa deposited in the female tract at mating or artificial insemination can be recovered from the Fallopian tubes. However, if a sufficient volume of semen is inseminated so that fluid reaches the utero-tubal junction, the number of spermatozoa entering the tubes is directly related to the concentration of spermatozoa in the inseminate (24). Moreover, the concentration of sperm at the site of fertilization is probably a function of the number of spermatozoa persisting at the utero-tubal junction.

Some disagreement exists as to the efficiency of transport of live versus dead spermatozoa in the reproductive tract of oestrous gilts. Whilst it is apparent that dead cells suspended in an appropriate medium may reach the upper uterine horns under little numerical disadvantage when compared with live cells, negotia-

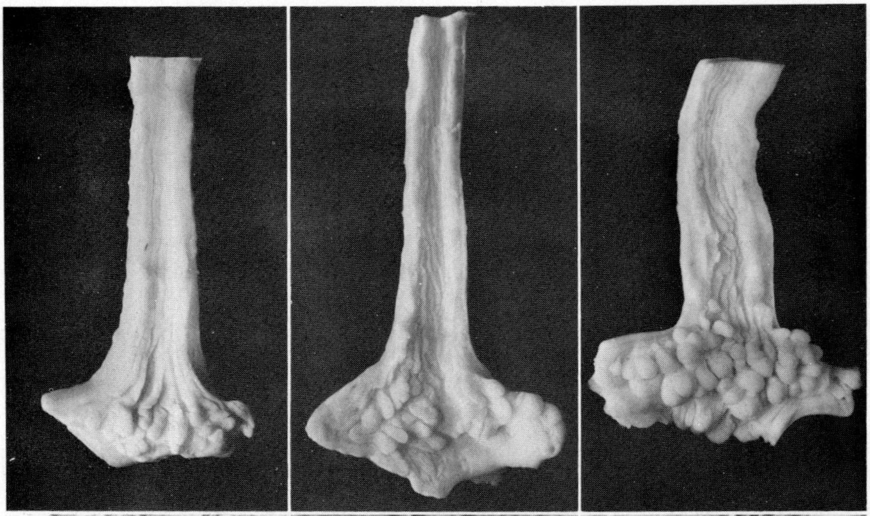

1, 2, 3

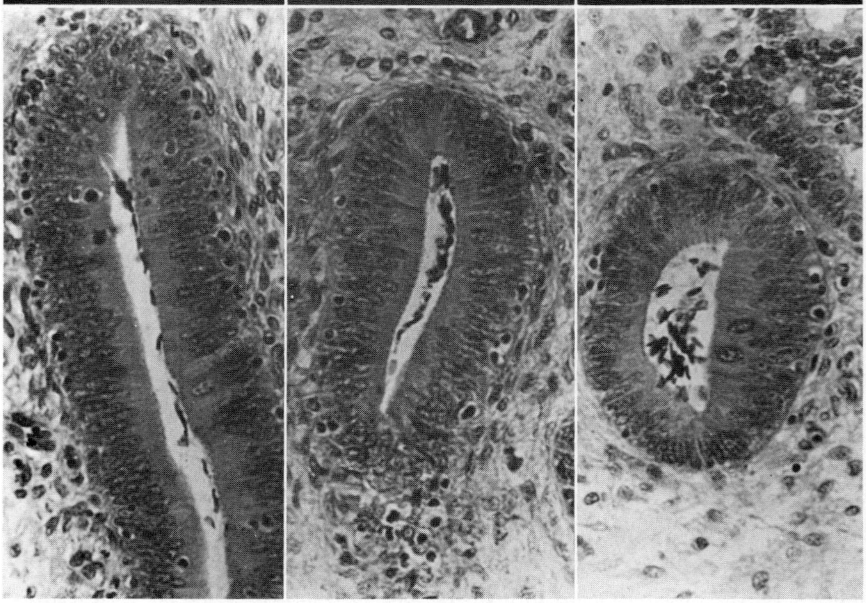

4, 5, 6

Fig. 1, 2, 3. Longitudinal sections through the porcine isthmus and utero-tubal junction at different stages of the oestrous cycle. Note the progressive increase in the swelling and oedematous condition of the polypoid processes at the utero-tubal junction in the three preparations. Figures 1 and 2 are of material obtained in the early follicular phase of the cycle, whereas that in figure 3 was taken from an oestrous pig shortly after the time of ovulation.

tion of the complex folds and processes of the utero-tubal junction (fig. 1–3) and entry into the isthmus are likely to be accomplished more efficiently by motile spermatozoa. Nevertheless, the degree of selection between live and dead cells in this region requires further study, not least to clarify if there is some form of selection against frozen-thawed boar sperm (23).

Although the utero-tubal junction may thus impede the rapid transport of large numbers of spermatozoa from the uterus to the site of fertilization, it has also been reported to act as a 'reservoir' from which spermatozoa gradually proceed to the isthmus (12, 24). But there is good evidence to suggest that the isthmus of the Fallopian tube also functions as a physiological sperm reservoir in the pig, as well as exerting further quantitative regulation over the population of spermatozoa ascending to the ampulla, thereby reducing the risk of polyspermic fertilization.

Boar spermatozoa do not remain alive in the vagina or cervix for more than 5 h, but they can survive elsewhere in the female tract for 24–48 h, or as long as 70–72 h in the region of the utero-tubal junction and terminal portions of the lower isthmus (fig. 4–6). In the uterine horns themselves, not only does the large volume of semen disappear within about 2 h, but sperm motility is drastically reduced and the spermatozoa are not viable for more than 30 h (12).

A significant loss of spermatozoa after mating occurs by means of voiding of semen through the cervical canal. The principal mechanism of removal for those spermatozoa not initially disposed of in this manner appears to be phagocytosis, numerous authors having presented observations on the ingestion of spermatozoa by polymorphonuclear leucocytes. Phagocytosis of sperm cells in the uterus may begin within 6–8 h of mating (21), but *Lovell and Getty* (19) have reported infiltration of leucocytes into the lumen some 30 min after service. Spermatozoa that have entered the Fallopian tubes are subject to a different fate. Apart from those that attach to or penetrate the zona pellucida, or enter cells of the corona radiata, spermatozoa may pass from the fimbriated end of the tube into the peritoneal cavity and in a retrograde manner from the isthmus into the uterus (unpublished observations). Sperm incorporation into the tubal mucosa has been documented in other species, but evidence for such a process in the pig has yet to be presented. Also unresolved remains the question of whether spermatozoa are phagocytosed by leucocytes in the Fallopian tubes, particularly in the 46–48 h after ovulation during which the embryos or unfertilized eggs reside there.

Fig. 4, 5, 6. Sections through the folds and diverticulae of the utero-tubal junction of a gilt killed approximately 20 h after mating. Spermatozoa appear to congregate in relatively high numbers in this region of the female tract. Those sperm in the diverticulae that project towards the isthmus may be protected from the post-coital invasion of leucocytes.

Acknowledgements

Unpublished experiments reported in this paper were supported by a grant from the Agricultural Research Council, for which grateful acknowledgement is made. I am indebted to a number of colleagues in Edinburgh for kindly commenting on portions of the manuscript, and to Mr. *G. Finnie* for printing the photographs.

References

1 *Austin, C.R.:* The 'capacitation' of the mammalian sperm. Nature, Lond. *170:* 326 (1952).

2 *Austin, C.R.:* Fate of spermatozoa in the uterus of the mouse and rat. J. Endocrin. *14:* 335–342 (1957).

3 *Austin, C.R.:* Fate of spermatozoa in the female genital tract. J. Reprod. Fertil. *1:* 151–156 (1960).

4 *Blandau, R.J.:* On the factors involved in sperm transport through the cervix uteri of the albino rat. Amer. J. Anat. *77:* 253–272 (1945).

5 *Blandau, R.J. and Money, W.L.:* Observations on the rate of transport of spermatozoa in the female genital tract of the rat. Anat. Rec. *90:* 255–260 (1944).

6 *Blandau, R.J. and Odor, D.L.:* The total number of spermatozoa reaching various segments of the reproductive tract in the female albino rat at intervals after insemination. Anat. Rec. *103:* 93–110 (1949).

7 *Braden, A.W.H. and Gluecksohn-Waelsch, S.:* Further studies of the effect of the T-locus in the house mouse on male fertility. J. exp. Zool. *138:* 431–452 (1958).

8 *Burger, J.F.:* Sex physiology of pigs. Onderstepoort J. vet. Res. Suppl. No. 2 (1952).

9 *Burkhardt, J.:* Sperm survival in the genital tract of the mare. J. agric. Sci. *39:* 201–203 (1949).

10 *Day, F.T.:* Survival of spermatozoa in the genital tract of the mare. J. agric. Sci. *32:* 108–111 (1942).

11 *Doak, R.L.; Hall, A., and Dale, H.E.:* Longevity of spermatozoa in the reproductive tract of the bitch. J. Reprod. Fertil. *13:* 51–58 (1967).

12 *Mesnil du Buisson, F. du et Dauzier, L.:* Distribution et résorption du sperme dans le tractus génital de la truie: survie des spermatozoïdes. Ann. Endocrin., Paris *16:* 413–422 (1955).

13 *Evans, E.I.:* The transport of spermatozoa in the dog. Amer. J. Physiol. *105:* 287–293 (1933).

14 *Griffiths, W.F.B. and Amoroso, E.C.:* Pro-oestrus, oestrus, ovulation and mating in the greyhound bitch. Vet. Rec. *51:* 1279–1284 (1939).

15 *Hancock, J.L. and Rowlands, I.W.:* The physiology of reproduction in the dog. Vet. Rec. *61:* 771–776 (1949).

16 *Hunter, R.H.F.:* Transport, migration and survival of spermatozoa in the female genital tract: species with intra-uterine deposition of semen; in *Thibault and Hafez* Transport, survie et pouvoir fécondant des spermatozoïdes chez les vertébrés (INSERM, Paris 1973).

17 *Hunter, R.H.F.:* Physiological aspects of sperm transport, migration and storage in the domestic pig, *Sus scrofa.* Brit. vet. J. (in press, 1975).

18 *Leonard, S.L. and Perlman, P.L.:* Conditions effecting the passage of spermatozoa through the utero-tubal junction of the rat. Anat. Rec. *104:* 89–102 (1949).

19 *Lovell, J.E. and Getty, R.:* Fate of semen in the uterus of the sow: histologic study of

endometrium during the 27 hours after natural service. Amer. J. vet. Res. *29:* 609–625 (1968).

20 *Mann, T.; Leone, E., and Polge, C.:* The composition of the stallion's semen. J. Endocrin. *13:* 279–290 (1956).

21 *Pitkjanen, I.G.:* The fate of spermatozoa in the uterus of the sow. Z. Obsc. Biol. *21:* 28–33 (1960).

22 *Polge, C.:* Artificial insemination in pigs. Vet. Rec. *68:* 62–76 (1956).

23 *Polge, C.; Salamon, S., and Wilmut, I.:* Fertilizing capacity of frozen boar semen following surgical insemination. Vet. Rec. *87:* 424–428 (1970).

24 *Rigby, J.P.:* The fate of spermatozoa in the genital tract of the sow following artificial insemination. Proc. 5th Int. Congr. Anim. Reprod., vol. 4, pp. 421–425 (1964).

25 *Rodolfo, A.:* The physiology of reproduction in swine. The semen of boars under different intensivenesses of mating. Philipp. J. Sci. *53:* 183–203 (1934).

26 *Soderwall, A.L. and Blandau, R.J.:* The duration of the fertilizing capacity of spermatozoa in the female genital tract of the rat. J. exp. Zool. *88:* 55–64 (1941).

27 *Thibault, C. and Hafez, E.S.E.:* Transport, survie et pouvoir fécondant des spermatozoïdes chez les vertébrés (INSERM, Paris 1973).

28 *Ventura, W.P. and Freund, M.:* Evidence for a new class of uterine stimulants in rat semen and male accessory gland secretions. J. Reprod. Fertil. *33:* 507–511 (1973).

29 *Yochem, D.E.:* Spermatozoon life in the female reproductive tract of the guinea pig and rat. Biol. Bull. *56:* 274–297 (1929).

Author's address: Prof. *R.H.F. Hunter,* School of Agriculture, University of Edinburgh, West Mains Road, *Edinburgh 9* (Scotland)

The Biology of Spermatozoa
INSERM Int. Symp., Nouzilly 1973, pp. 156–165 (Karger, Basel 1975)

Transport and Survival of Spermatozoa in Cattle

C. Thibault, Micheline Gerard and Y. Heyman

Université de Paris VI and Station centrale de Physiologie animale INRA, Jouy-en-Josas

Introduction

The duration of sperm transport in the female tract of mammals, between the vagina and the ampullary portions of the Fallopian tube, and how long fertilizing ability can be maintained are questions which have received controversial answers.

Exact answers have practical significance in domestic mammals where artificial insemination is now routinely employed on a large scale.

In cattle, *van Demark and Moeller* (12), *van Demark and Hays* (11) have recovered spermatozoa in ampulla by perfusion 2 min 30 sec after mating. In fact, this time represents the minimum delay between coitus and slaughter. *Brewster et al.* (1), using the same technique, concluded that the minimum duration is 4 h 25 min in heifers, and 5 h 30 min in cows. *Dauzier and du Mesnil du Buisson* (2), using smears of the internal sides of isthmus and ampulla frag-

Table I. Duration of sperm transport in cattle. Presence of spermatozoa in the Fallopian tube

	Techniques of control	Authors
Short duration		
3 min 15 sec	perfusion	*Van Demark and Moeller* (12)
2 min	perfusion	*Van Demark and Hays* (11)
Long duration		
4 h 25 min to 5 h 30 min	perfusion	*Brewster et al.* (1)
3–6 h	perfusion	*Sheikin* (1972)
8 h	smears and perfusion (ampulla)	*Dauzier and du Mesnil du Buisson* (2)
12 h	postovulatory fertilization	*Thibault* (8)

ments, were unable to find spermatozoa before 8 h. *Thibault* (8) did not observe the first steps of fertilization before 12 h after insemination, performed after ovulation had occurred (table I).

It seems impossible to explain these contradictory results, 2 min to 12 h, by physiological hypotheses. Thus, it appears necessary to end this controversy by using an irreproachable technique, that of microscopic examination of serial sections of genital tract fixed as a whole after slaughter. Moreover, this technique gives a complete view of the dynamic distribution of sperm along the tract at increasing times after coitus.

Material and Methods

Ten Frisian cows of proven fertility were mated once during estrus. Estrus was detected twice a day by vasectomized bulls. Mating generally occurred during the first half of estrus. Two cows for each series were slaughtered in the laboratory slaughterhouse 2, 8, 18, 48, 72 h after mating. Immediately after death, the whole genital tract was collected by opening the abdominal wall. The uterine horns and Fallopian tubes were quickly dissected, stretched and tied to a glass plate, and then immersed in a fixative bath (Bouin-Holland) for 10–15 days. Seven successive segments, of approximately 2 cm length each, were taken between the utero-tubal junction and the ampulla. These were embedded in paraffin. Longitudinal serial sections of 13 μm thickness were stained with hematoxylin-eosin. Spermatozoa were searched for under the microscope.

The exact site of each sperm was recorded so that, after examination of all serial sections, real distribution of spermatozoa in parts of the female tract could be reconstituted.

Times from mating to slaughter were chosen in order to determine if fast ascent occurred (2 h), and how the number of sperm present increased up to the time of ovulation (8 and 18 h). We wished to know how long normal, and apparently living, spermatozoa remained in the female tract at the utero-tubal junction at precisely 48 and 72 h after mating.

In order to study the relation between estrus and sperm survival, sperms were inseminated directly into the uterine horn by vagino-transperitoneal route, according to *Testart and Leglise* (7). In this way, it is possible to study the influence of the uterine milieu without interference from cervical permeability. Twelve cows were inseminated with 20×10^6 to 600×10^6 spermatozoa from bulls of proven fertility, either during estrus or 24–48 h before. All animals were slaughtered 72 h after mating, and sperm counts were made using the same procedure as before.

Results

Duration of Sperm Transport after Natural Mating

Microscopic examination of the genital tract of two cows slaughtered 2 h after coitus (cows 223 and 251) showed that very few spermatozoa had reached the utero-tubal junctions: 2, 63 and 106, respectively, were found in three UTJ

Table II. Sperm distribution from UTJ to ampulla at different intervals of time after coitus

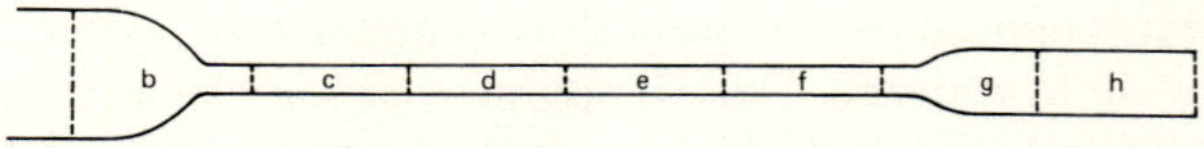

Time after coitus	UTJ	Isthmus				Ampulla		Cow No.
	b	c	d	e	f	g	h	
2 hours								
Left tube	108	–	0					223
Right tube	63	6	0					
Left tube	2	0	–					251
Right tube	–	0	–					
8 hours								
Left tube		208	36	20	16	6	0	252
Right tube		229	195	3	3	48	2	
Left tube	–	–	–	–	–	–		241
Right tube	3,857	49	2	–	–	–		
18 hours								
Left tube	20,000	1,138	533	686	20	3	3	236
Right tube	–	358	819	227	5	0	–	
Left tube	22,842	–	–	–	–	–	10	259
Right tube	20,844	–	–	–	–	–	–	
48 hours								
Left tube	1,327	42	53	–	–	16	–	277
Right tube	2,053	204	43	0	–	–	–	
Left tube	94	165	114	100	4	12	–	273
Right tube								
72 hours								
Left tube	–	126	8	133	1,264	–	–	341
Right tube	1,788	93	5	2	0	3	183	
Left tube	–	2	1	–				353
Right tube	179	4	1	–				

studied (table II). In the first 2-cm segment of the lower isthmus, no sperms were present in two segments, and only six in the third. In this case, the six spermatozoa represented the sperm ascent front, since no sperms had been found in the next 2-cm segment (cow 223).

Thus, there was no rapid ascent of spermatozoa in the Fallopian tube in the cow after mating. These results were similar to those found in unstressed ewes (10).

As the time between mating and slaughter increased, the number of spermatozoa in the Fallopian tube and UTJ became higher.

Eight hours after mating, sperm, numbering up to many thousands, were found in the UTJ, and in one of the two cows, had reached the ampulla. However, the total number of sperms present in the isthmus and the beginning of the ampulla was limited to several hundreds.

It appears at this moment that the UTJ acts as a storage container, and the lower isthmus as a filter, narrowly restricting the number of spermatozoa able to attain the site of fertilization (table II).

This double function of the UTJ and lower isthmus was even more patent 18 h after coitus (table II). Sperm countings in the UTJ of the two cows studied gave approximately 20,000 sperms for each UTJ. The total number of sperms in the isthmus was also much higher than at 8 h (compare cows 252 and 236, table II), but the same filter-like action remained evident since, near the upper part of the isthmus, there was an abrupt decrease from some hundreds of sperms per each 2-cm segment to less than 10 for the superior segments.

Sperm Distribution

Utero-Tubal Junction

As soon as spermatozoa arrive in the uterus, some of them migrate inside the uterine glands (fig. 2a). At the level of the utero-tubal junction, uterine glands are replaced by uncoiled folds. They are conical-shaped and run almost parallel to the lumen of the genital tract. Spermatozoa penetrate there in great numbers.

At increasing times after coitus, the proportion of spermatozoa inside the glands and folds becomes higher (table III).

Forty-eight and 72 h after coitus, spermatozoa are rarely present in the lumen (table III). Those which have penetrated the glands are disappearing, destroyed by macrophages (fig. 2b). Only those found in some UTJ folds are alive (fig. 2c, d).

Isthmus-Ampulla

Eight and 18 h after coitus, the number of spermatozoa in the isthmus decreases progressively from the UTJ to the ampulla (table II).

On the other hand, from 18 h, spermatozoa are unequally distributed along the Fallopian tube; groups of spermatozoa are separated by areas where spermatozoa are practically missing (fig. 1, 18 h, cow 236; table II, 72 h, cow 341).

Formation of bundles of spermatozoa is a result of local, muscular contractions, isolating relatively distended segments, as clearly demonstrated by *Wintenberger-Torres* (13) on the sheep isthmus; spermatozoa would be pushed and maintained in bulges. Dead and degenerating sperms are found beside morphologically normal spermatozoa in these bundles 48–72 h after mating.

Table III. Proportions of spermatozoa in uterine glands and tubal folds at UTJ level

Cow No.	Intervals from mating to slaughtering h	Spermatozoa in	
		lumen	uterine glands or folds
223	2	82	26
		41	22
252	8	366	497
241	8	2,605	1,252
159	18	8,328	12,516
273	48	10	84
277	48	26	1,301
341	72	28	1,760
353	72	4	175

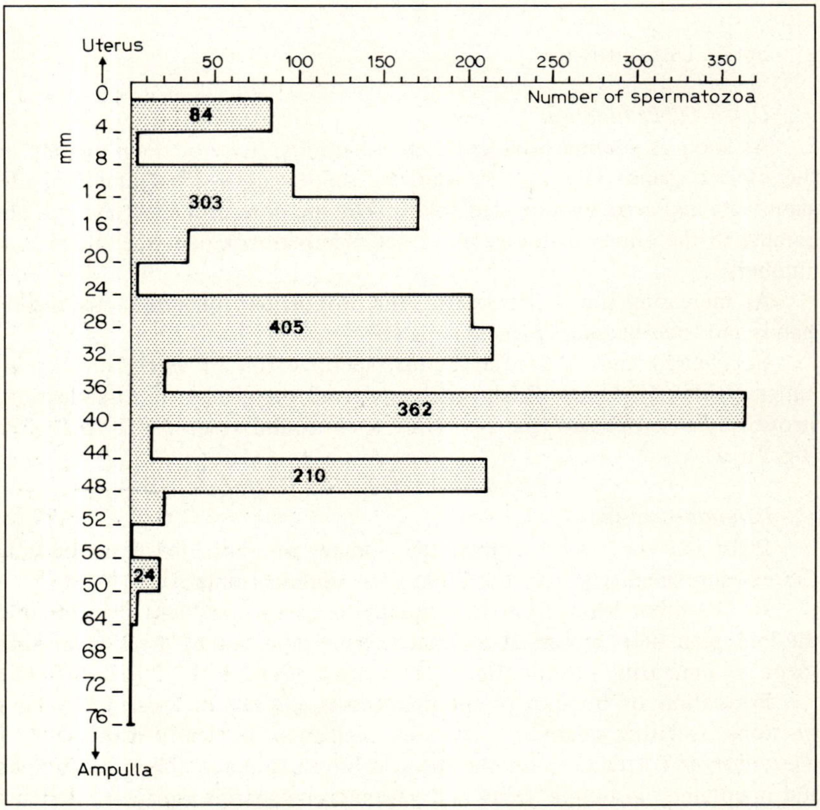

Fig. 1. Sperm distribution along 7.6 cm of isthmus in cow 236, 18 h after mating.

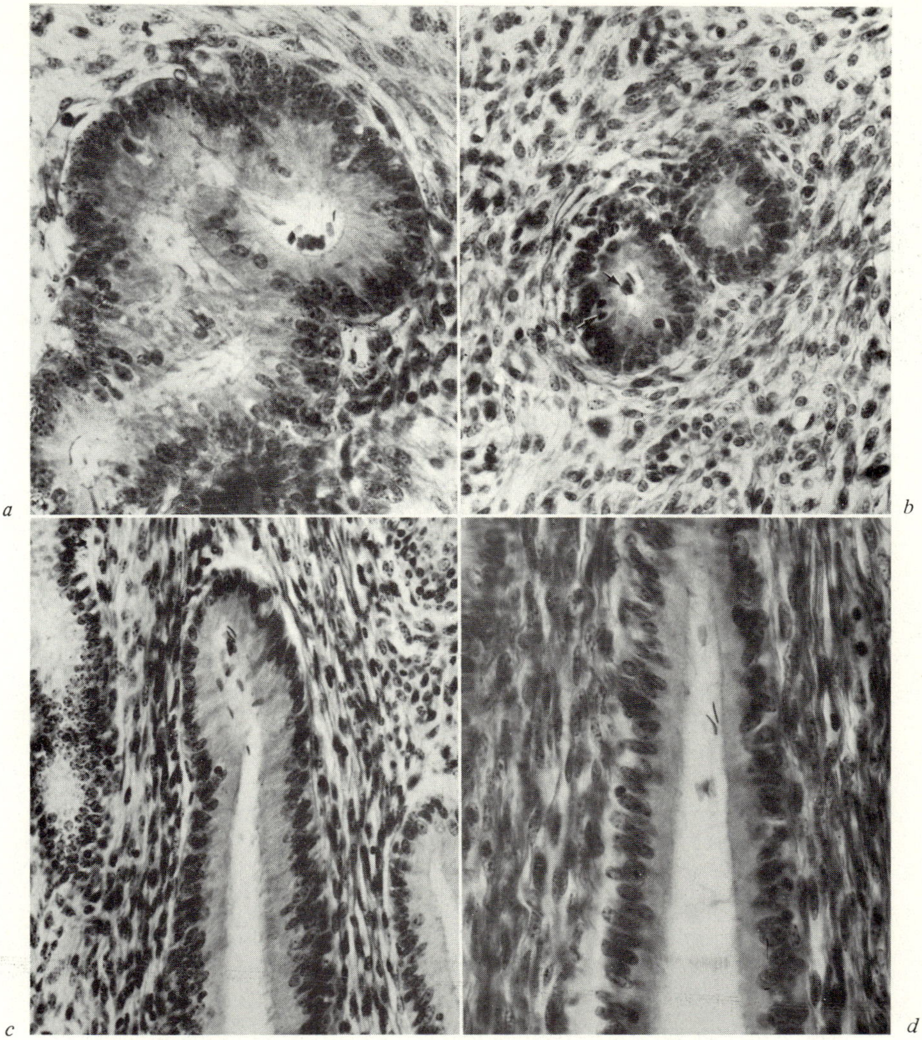

Fig. 2. Spermatozoa in uterine glands or folds. *a* Spermatozoa inside the uterine glands 18 h after coitus. *b* Spermatozoa phagocytized by leukocytes invading uterine glands 48 h after mating. Note other leukocytes penetrating the gland. *c* and *d* Spermatozoa inside the UTJ folds 72 h after coitus. Spermatozoa resemble living sperms and lymphocytes are absent.

Sperm Preservation

Because of the technique used, it is impossible to assume that spermatozoa are living or dead at slaughter time. However, dead spermatozoa are easily recognizable by the sharp outline of the head which becomes smooth and swells slightly. Also, the head and tail are often detached. This was the case of many spermatozoa in the lumen of the isthmus or UTJ 48 or 72 h after coitus, but in UTJ folds, spermatozoa were twisted, as living sperms, and had wavy tails.

Natural Mating

According to these criteria, living spermatozoa are always present at 48 and 72 h in the UTJ folds, while they progressively disappear in the uterine glands after 18 h.

This difference seems to result from the absence of leukocytes in the folds, whereas they invade uterine glands and stroma (fig. 2b, d).

There are individual differences in the importance of leukocytic invasion. An invasion appears more important in cows 273 and 353, which had only a few sperms in UTJ segment, than in cows 277 and 341 where numerous sperms remained in the same segment, and in the isthmus (table II).

Forty-eight and 72 h after coitus, isolated sperms in the lumen of the isthmus generally have a normal appearance. However, those in bundles are often included in mucus where dead sperms or other cellular debris are also visible. Thus, it seems reasonable to suppose that, even while normal, they are also dying.

The relatively long survival of bull sperm in the UTJ folds occurs after natural mating. Spermatozoa have reached this site at the end of estrogenic phase and their survival is observed during the beginning of the luteal phase.

Intrauterine Insemination

If spermatozoa are deposited directly inside the uterine lumen by transperitoneal route, in order to avoid variation of cervix permeability before estrus, the frequency of survival at UTJ, 72 h later, is lower than insemiation which has been done 24 h before estrus, and is nil for a 48-hour delay (table IV).

The ability of sperms to migrate and survive in the UTJ folds in dependent on the ovarian steroid secretion levels.

It was noted that in two cows (296 and 365), where insemination was limited to the left uterine horn, spermatozoa had been found in the two UTJ, although the number in the right UTJ was lower.

Finally, when individual differences were evident at the UTJ and isthmic levels, no relationship existed between the number of sperm inseminated and the number present 72 h later. At least, this was found in the limit of the numbers used, from 20×10^6 to 600×10^6 (table V).

Table IV. Presence (+) or absence (–) of spermatozoa at UTJ level 72 h after intrauterine insemination

Time of insemination	Cow No.	Sperm at UTJ level
48 h before estrus	262	–
	364	–
	369	–
24 h before estrus	270	–
	362	–
	226	+
	326	–
	296	+
	313	–
	368	–
During estrus	394	+
	365	+

Table V. Number of sperms at UTJ level 72 h after intrauterine insemination

Cow No.	Number of sperm inseminated $\times 10^6$	Number of sperms	
		in left UTJ	in right UTJ
226	600	360	289
394	340	89	70
296	20	47	21
365	20	373	80
Control			
353	natural mating		179
343	natural mating		1,788

Discussion

The discussion shall be limited to three particular aspects of this research:

a) No fast ascent of spermatozoa after mating. Using our technique, post mortem contamination by outside spermatozoa is absolutely impossible. *Van Demark*'s conclusions must be discarded as erroneous, probably because of contamination caused by the use of a warm bath to which the clamped genital tract was maintained during perfusion.

The same opposite conclusions can be found in the literature on the ewe (9). Succeeding publications have only served to enlarge these two factions. For example, *Lang and Oh* (4) concluded a 4- to 16-hour transport, while *Lightfoot and Restall* (5) found spermatozoa in the Fallopian tubes 30 min after insemination with live or dead sperms in 13–25 % of ewes.

Using the same technique in the ewe as in the cow when studying serial sections after fixation *in toto* of the genital tract, we noticed that the first spermatozoa never reached the isthmus before 2 h (10). These experiments took into account the unfavorable effect of stress due to manipulation of the animal (6), since ewes were mated freely at pasture.

b) The total number of spermatozoa present in the isthmus and beginning of the ampulla remained limited to several thousands, even 18 h after mating, at which time the maximum number of spermatozoa were found. If we include spermatozoa present in the UTJ and beginning of the isthmus (B segment), then about 25,000 sperms were present in each Fallopian tube of the cow. This number was similar to that indicated by *El Banna and Hafez* (3) who found an average of 17,000 spermatozoa 16 h after mating.

c) The presence of finger-shaped folds at the UTJ in place of coiled uterine glands is of peculiar interest since spermatozoa can survive at least three days in these structures. It seems that this relatively long-term survival results from the absence of leukocytes in the folds when the lower part of the genital tract is cleaned more or less rapidly, depending on the animal.

References

1 *Brewster, J.E.; May, R., and Cole, C.A.:* The time of ovulation and rate of sperm travel in cattle. Proc. 33rd Ann. Meet. Amer. Soc. Anim. Prod., pp. 304–310 (1940).

2 *Dauzier, L. et Mesnil du Buisson, F. du:* Les spermatozoïdes dans l'appareil génital femelle (étude chez la brebis, la vache, la truie). Proc. 2nd Wld Congr. Fertil. Steril., Naples 1956, pp. 1149–1159.

3 *Banna, A.A. El and Hafez, E.S.E.:* Sperm transport and distribution in rabbit and cattle female tract. Fertil. Steril. *21:* 534–540 (1970).

4 *Lang, D.R. and Oh, Y.K.:* Distribution of spermatozoa in the reproductive tract of the Romney ewe. New Zeald J. agric. Res. *13:* 157–166 (1970).

5 *Lightfoot, R.J. and Restall, B.J.:* Effects of site of insemination, sperm motility and genital tract contractions on transport of spermatozoa in the ewe. J. Reprod. Fertil. *26:* 1–13 (1971).

6 *Mattner, P.E. and Braden, A.W.H.:* Spermatozoa in the genital tract of the ewe. I. Rapidity of transport. Austr. J. Biol. Sci. *16:* 473–481 (1963).

7 *Testart, J. et Léglise, P.C.:* Transplantation d'œufs divisés chez la vache, par voie transvaginale. C.R. Acad. Sci., D *272:* 2591–2592 (1971).

8 *Thibault, C.:* Analyse comparée de la fécondation et de ses anomalies chez la brebis, la vache et la lapine. Ann. Biol. anim. Biochem. Biophys. *7:* 5–23 (1967).

9 *Thibault, C.:* La fécondation chez les mammifères; in *Grassé* Traité de zoologie, tome 16, fasc. 6, pp. 911–963 (Masson, Paris 1969).

10 *Thibault, C. and Wintenberger-Torres, S.:* Oxytocin and sperm transport in the ewe. Int. J. Fertil. *12:* 410–415 (1967).
11 *Demark, N.L. van and Hays, R.L.:* Rapid sperm transport in the cow. Fertil. Steril. *5:* 131–137 (1954).
12 *Demark, N.L. van and Moeller, A.M.:* Speed of spermatozoan transport in the reproductive tract of oestrous cow. Amer. J. Physiol. *165:* 674–678 (1951).
13 *Wintenberger-Torres, S.:* Mouvements des trompes et progression des œufs chez la brebis. Ann. Biol. anim. Biochem. Biophys. *1:* 121–133 (1961).

Author's address: Prof. *C.G. Thibault,* Station de Physiologie animale, Institut National de la Recherche Agronomique, *F–78350 Jouy-en-Josas* (France)

The Biology of Spermatozoa
INSERM Int. Symp., Nouzilly 1973, pp. 166–173 (Karger, Basel 1975)

Effect of Artificial Insemination with Small Numbers of Frozen Sperm on Fertility and Sperm Transport in Cattle

R. Jondet

Union Régionale des Coopératives d'Insémination Artificielle, Rennes

One of the principal advantages of artificial insemination in domestic animals is that a single ejaculation can be diluted and subdivided for use among a large number of females. The history of bovine semen dilution, or more precisely, the history of maximal use of collected spermatozoa, closely follows that of sperm technology and it is convenient to distinguish three principal stages: (1) Use of undiluted semen (1935–1943), in which an ejaculation was divided into several fractions which were immediately inserted into an equal number of cows either in the vagina (gelatine capsules) or in the cervix (using a speculum). The volume of the dose varied from 0.1 to 4.0 ml and the number of spermatozoa was at least 150×10^6. (2) Use of diluted semen stored at 5 °C, in which the pioneer efforts of *Anderson* (1) in 1941 and *Salisbury et al.* (2) in 1943 provided the starting points for numerous experiments on varying sperm concentrations per insemination dose. From 1943 to 1972, the typical total sperm number per dose decreased from 74×10^6 to 0.5×10^6. (3) Use of frozen semen, in which the originators of the process in 1952, *Polge and Rowson* (3), only used a very low semen dilution rate (1:3). As techniques improved, however, workers became bolder and prepared a larger and larger number of doses from the same quantity of neat semen or total spermatozoa in the ejaculation. The minimal number of spermatozoa per dose (based on those showing progressive motility after thawing) was reduced from about 5×10^6 in 1953 to 0.375×10^6 in 1972.

This report concerns the effect of inseminating relatively low numbers of motile spermatozoa upon fertility and rate of sperm transport in the bovine.

Table I. Influence of freezing and thawing on survival rate of spermatozoa: comparison of 3 concentrations (means for 4 bulls)

Motile sperm number per dose											
13×10^6			3×10^6			0.375×10^6			means		
BF	AT	loss	BF	AT	loss	BF	AT	loss	BF	AT	loss
68.75	65.75	3.00	68.75	65.00	3.75	68.75	63.75	5.00	68.75	64.83	3.92

BF = Before freezing; AT = after thawing.

Materials and Methods

Fertility Trial

One ejaculate from each of four Normand bulls with good fertility (74, 68, 71 and 68 % 60- to 90-day nonreturn rate) were used. Sperm concentration for the four ejaculates was 2,150, 1,503, 2,528 and $1{,}656 \times 10^6$ per ml. For each bull, three series of doses were prepared containing either 13×10^6, 3×10^6 or 0.375×10^6 progressively motile sperm after thawing. These three concentrations were selected on the basis of an expected post-thaw sperm survival rate of 65 % (this expectation was achieved as noted in table I). Techniques used for preparation and packaging of the semen, freezing of the straws (0.20 ml) in vertical position using N^2 vapor, and thawing were identical to those described previously (4, 5). The temperature decrease during freezing was exactly the same for all doses (5), so that this was not a variable in the results obtained using these very small numbers of spermatozoa. The frozen semen was used by 12 technicians, selected because of their high efficiency, to inseminate 2,988 first service cows, most of which had calved at least 60 days before insemination.

Sperm Transport

Three cows were inseminated (deep intracervical) with a semen dose containing 0.375×10^6 progressively motile spermatozoa. At 16, 18 or 72 h following insemination, one cow was slaughtered and the uterus and oviducts removed and fixed. The spermatozoa were detected and counted on 20-mm serial longitudinal sections from the uterotubal junction (UTJ) and first segments of the isthmus (6).

Results and Discussion

Fertility Results

Table II shows that even though use of the lowest number of spermatozoa (0.375×10^6) yielded a noticeably lower conception rate than the two higher concentrations it still yielded a mean 60- to 90-day nonreturn rate of 51.85 %. These results agree with those we previously obtained (4).

Table II. Influence of reduction in motile sperm number per insemination (doses of 0.20 ml) on conception rate

Bulls	Motile sperm number per dose (0.20 ml)								
	13×10^6 (control)			3×10^6			0.375×10^6		
	number of cows	NR 60–90 days	% NR 60–90 days	number of cows	NR 60–90 days	% NR 60–90 days	number of cows	NR 60–90 days	% NR 60–90 days
A	251	180	71.71	246	151	61.38	258	149	57.75
B	254	173	68.11	237	139	58.64	216	97	44.90
C	262	189	72.13	261	168	64.36	234	134	57.26
D	245	161	65.71	262	142	54.19	262	123	46.94
Means	1,012	703	69.46[1]	1,006	600	59.64[1]	970	503	51.85[1]

1 Difference statistically significant at the 5-percent level.

Table III. Insemination with a reduced number of motile spermatozoa

Bulls	Fertility level	13×10^6 motile sperm			3×10^6 motile sperm			0.375×10^6 motile sperm		
		number of cows	NR 60–90 days	% NR 60–90 days	number of cows	NR 60–90 days	% NR 60–90 days	number of cows	NR 60–90 days	% NR 60–90 days
A + C	73.32	513	369	71.92	507	319	62.91	492	283	57.62
B + D	68.57	499	334	66.93	499	281	56.31	478	220	46.02
				–4.99*			–6.60**			–11.50***
Means	70.51	1,012	703	69.46	1,006	600	59.64	970	503	51.85

Influence of bull fertility level on conception: * difference not significant; ** difference significant; *** difference highly significant at the 5-percent level.

The conception rate obtained with a small number of sperm is influenced by several factors, notably the fertility level of the bull, volume of the insemination dose, time of insemination, freezing method and technician efficiency.

Effect of the Fertility Level of the Bull

Sullivan and Elliott (7) showed that when sperm numbers per insemination are increased, conception rates are increased but to a lesser degree for the least fertile bulls. As shown in table III, in this study the most fertile bulls were those whose semen was particularly well suited to high dilution rate (57.52 % non-return rates with 0.375×10^6 spermatozoa for the two most fertile bulls against 46.02 % for the other two). These results confirm our earlier observations.

Thus high dilution rates yield best fertility results when applied to high quality ejaculates collected from bulls of relatively high fertility. Bulls A and C in table II which gave the best fertility results with a dose of 0.375×10^6 motile sperm, were also those whose ejaculations contained the largest numbers of motile spermatozoa. Experience indicates that conception rates achieved with inferior quality ejaculations cannot be markedly improved solely by increasing the number of sperm per insemination dose. It is also known that diluent composition affects fertility. For example, addition of fructose (8) and antibiotics (9–11) to diluted semen usually enhances conception rate.

Effect of the Volume of Insemination Dose

For a long time the volume remained at about 1.0 ml per insemination. This volume has been reduced progressively in french straws to 0.45 and 0.2 ml (12). The reduction from 13 to 6.5 and 3.25×10^6 spermatozoa in 0.2 ml has not altered conception rate significantly. Doses of 0.1 ml, however, cause a noticeable decrease in conception rate with 3.25×10^6 spermatozoa (13). In addition to the greater inconvenience of handling such a small dose, it appears likely that the decrease in fertility with 0.1 ml would be much greater if even fewer sperm had been used (0.375×10^6 rather than 3.25×10^6).

Effect of Insemination Time

Insemination time may have a direct influence on speed of transport as well as on sperm survival. For cows slaughtered 16 and 40 h after cervical insemination (with a speculum), the number of sperm present in the oviducts was four times higher for 16 h than for 40 h (14). Ovulation in the cow occurs an average of 11 h after the end of estrus. If inseminated at the proper time toward the end of estrus, the number of spermatozoa at the site of fertilization in the upper portions of the oviducts should peak at ovulation. The oviducts must be constantly resupplied with spermatozoa from the inferior regions (UTJ and cervix) which act as reservoirs (15). Recent observations indicate that only insemination during estrus allows the recuperation of spermatozoa at the UTJ at 72 h. Moreover, the number of cells counted in the oviducts of some tens or hundreds only (16), bears no relationship to the quantity inseminated (17).

Sperm counts made on the genital tracts of the three experimental cows that were inseminated with 0.375×10^6 motile spermatozoa indicate that the number of cells found in UTJ and the lower part of the isthmus is very low (table IV). The fertility results obtained reveal a sharp fall after the 24th hour (fig. 1). This finding may be due to the fact that, as the time from the end of

Table IV. Number and distribution of spermatozoa at utero tubal junction (UTJ) and in the two first segments of the isthmus, after insemination with 0.375×10^6 motile sperm

Idendity number of cow	Interval between insemination and slanghter, h	Number of spermatozoa observed				
		UTJ (20 mm)			isthmus 1 (20 mm)	isthmus 2 (20 mm)
		lumen	folds	total		
402	16	9	12	21	0	1
309	72	0	0	0	0	–
377	18	65	65	130	25	–

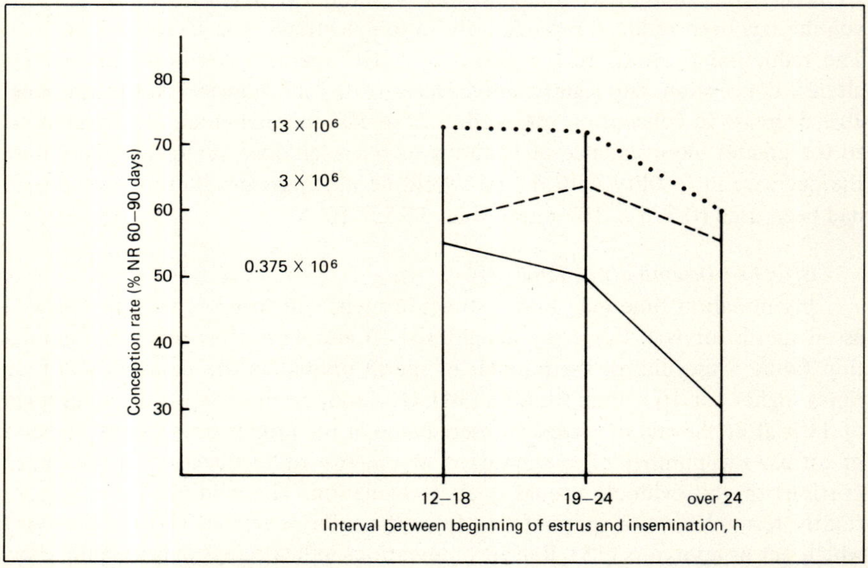

Fig. 1. Effect of insemination time with a reduced sperm number on conception rate.

estrus increases, the female genital organs become less and less favorable for sperm survival. A massive destruction of cells then follows which take on particular importance when a very low number of sperm are inseminated.

Effect of the Freezing Method

One cannot exclude the possibility that the method of freezing may affect the fetilizing capacity of frozen spermatozoa. The fertility and transport results reported here were obtained with a freezing technique which kills not more than 5 % of sperm. With a method more injurious to spermatozoa, one may think that these results could be different.

Effect of Technician Efficiency

Technician efficiency and the means of improving it have received much attention from research workers and managers of artificial insemination units (18, 19). It has not been possible for us to provide information on this subject because we only retain technicians with high efficiency. A study undertaken in Great Britain, however, showed that reducing sperm numbers from 14 to 7×10^6 had no effect on conception rates obtained by first grade technicians while conception rates decreased by 10 % for second grade technicians (20). Consequently, semen doses containing still smaller numbers (20 times less) might give good results only in the hands of highly competent technicians.

Sperm Transport

Table IV shows that the number of sperm present at UTJ and in the first segments of the isthmus is very small, 18 h post insemination. If the delay is exteded to 72 h, no spermatozoa are found although, normally, following the work of *Thibault et al.* (16), these figures, at 18 h, reach about 20,000 at the UTJ and several hundreds in the first segments of the isthmus. Even though those numbers were very low and there were no sperm present at 72 h, the eggs recovered, when ovulation occurred, were fertilized.

Conclusion

The results of this study confirm our earlier finding that it is possible to achieve fertility in cows inseminated with 0.375×10^6 motile spermatozoa. Although conception rate was noticeably lower than that obtained with 13×10^6 motile sperm, the level of fertility was sufficiently high to warrant consideration in situations involving maximum use of a genetically superior bull.

High semen dilution makes it possible to prepare a very large number of doses so that a large number of cows can be inseminated with sperm from a single ejaculation. This makes it possible to exploit greatly the economic impact of superior bulls which are likely to disappear from breeding service prematurely because of illness, a physical defect or some other reason.

It is surprising that placement of a number of spermatozoa as small as 0.375×10^6 in the rather large genital tract of the cow can achieve fertilization at the level reported herein. Experience indicates, however, that these results can only be obtained using semen of bulls of relatively high fertility, a freezing method which provides a high sperm survival rate after thawing, and highly competent technicians.

Acknowledgements

The author wishes to thank Prof. *Thibault* (INRA) for his kind help in the design of this work, Mr. *Courot,* Mr. *Heyman* and Mrs. *Gérard* (INRA) for valuable and much appreciated technical assistance, Miss *Solari* (INRA) for statistical analysis and Mr. *Rabadeux* (AI Unit of Rennes) for his loyal and efficient help.

References

1 *Anderson, J.:* Further investigations on artificial insemination of cattle. J. Agric. Sci. *31:* 348–352 (1941).

2 *Salisbury, G.W.; Beck, G.H.; Cupps, P.T., and Elliott, I.:* The effect of dilution rate on the livability and the fertility of bull spermatozoa used for artificial insemination. J. Dairy Sci. *26:* 1057–1069 (1943).

3 *Polge, C. and Rowson, L.E.A.:* Fertilizing capacity of bull spermatozoa after freezing at –79 °C. Nature, Lond. *169:* 626–628 (1952).

4 *Jondet, R.:* Contribution to the assessment of the minimal number of frozen spermatozoa necessary to obtain fertilization in the cow. 7e Congr. int. Reprod. Anim. Insém. Artif., Munich 1972, vol. 2, pp. 1445–1448.

5 *Jondet, R.:* Adaptation to 0.20 ml straws of the vertical freezing method in the liquid nitrogen vapour phase. 7e Congr. int. Reprod. Anim. Insém. Artif., Munich 1972, vol. 2, pp. 1350–1953.

6 *Thibault, C.:* Personal commun. (1973).

7 *Sullivan, J.J. and Elliott, F.I.:* Bull fertility as affected by an interaction between motile spermatozoa concentration and fertility level in artificial insemination. 6e Congr. int. Reprod. Anim. Insém. Artif., Paris 1968, vol. 2, pp. 1307–1310.

8 *Hafs, H.D. and Elliott, F.J.:* The effect of methods of adding egg yolk and monosaccharides on the survival of frozen bull spermatozoa. J. Dairy Sci. *38:* 811–815 (1955).

9 *Almquist, J.O.; Thorp, W.S.T., and Knodt, C.G.:* The effect of penicillin upon the livability, fertility and bacterial content of bull semen. J. anim. Sci. *5:* 400 (1946).

10 *Dunn, H.O.; Larson, G.L., and Willett, E.L.:* Preliminary breeding results with frozen semen. J. Dairy Sci. *36:* 578 (1953).

11 *Jondet, R.; Rabadeux, Y.; Pineau, R. et Jondet, M.:* Concentration du dilueur en antibiotiques et pouvoir fécondant du sperme congelé de taureau. Soc. Sci. Bretagne *46:* 267–271 (1971).

12 *Cassou, R.:* La miniaturisation des paillettes. 6e Congr. int. Reprod. Anim. Insém. Artif., Paris 1968, vol. 2, pp. 1009–1011.

13 *Jondet, R.:* Influence de la réduction du volume de la dose du sperme congelé et de sa teneur en spermatozoïdes sur la fécondation des vaches inséminées artificiellement. 6e Congr. int. Reprod. Anim. Insém. Artif., Paris 1968, vol. 2, pp. 1061–1068.

14 *Banna, A.A. El and Hafez, E.S.E.:* Sperm transport and distribution in rabbit and cattle female tract. Fertil. Steril. *21:* 534–540 (1970).

15 *Thibault, C. et Levasseur, M.C.:* Conservation et survie prolongée des spermatozoïdes dans les voies génitales femelles des vertébrés. Ann. Biol. anim. Biochem. Biophys. *13:* 267–284 (1973).

16 *Thibault, C.; Gerard, M. et Heyman, Y.:* Transport et survie des spermatozoïdes chez la vache. Symp., Nouzilly 1973.

17 *Heyman, Y.:* Survie des spermatozoïdes dans la jonction utéro-tubaire de la vache; D.E.A. Université Paris (1973).

18 *Pickett, B.W.:* Factors affecting the utilization of frozen bovine semen for maximum reproductive efficiency. AI Digest, Feb. 1971, p. 8.

19 *MacPherson, J.W.:* Semen placement effects on fertility in bovines. J. Dairy Sci. *51:* 807–808 (1968).

20 *Hunter, W.K.:* Glycerolization and freezing techniques with bull semen. 6e Congr. int. Reprod. Anim. Insém. Artif., Paris 1968, vol. 2, pp. 1053–1056.

Author's address: *R. Jondet,* Centre d'Insémination Artificielle, 63, rue de la Motte-Brûlon, *F–35000 Rennes* (France)

The Biology of Spermatozoa
INSERM Int. Symp., Nouzilly 1973, pp. 174–181 (Karger, Basel 1975)

Amino Constituents of Tubal and Uterine Fluids of the Estrous Ewe: Comparison with Blood Serum and Ram Seminal Fluid

Yves J.R. Menezo

INRA, Laboratoire de Biologie, INSA, Villeurbanne

Introduction

At estrous time female genital tract secretions constitute a necessary and adequate medium for the transition of gametes. These secretions contribute to the capacitation process put forward by *Mattner* (8) for ram spermatozoa. Recent data (10) show that in some species serum albumin is important for capacitation studied *in vitro*. Proteins of the female genital tract secretion were compared to those of vasectomized ram seminal plasma and blood plasma. Since capacitation, according to several authors (2, 17), could also be the result of neutralizing an inhibition, hydrolytic enzymes were tested in the same fluids using a semi-quantitative method. Free amino acids, which are important intermediary metabolites for nutrition and protein metabolism, were also studied. It was interesting to compare tubal and uterine secretions with the seminal plasma during sperm ascent time in the female tract. Plasma was analyzed comparatively to show the relations between the vascular compartment and the fluids secreted by the female and the male genital tracts.

Materials and Methods

Collection of Fluids

The estrous behavior of the Préalpes and Ile de France ewes was checked daily with a vasectomized ram.

Female Genital Tract Secretions

Oviductal secretions were collected by a cannulation technique according to *Perkins et al.* (11). A cannulation was made at the uterotubal junction to collect uterine fluids. Analyses were performed on 6 animals the first day of heat and the following day.

Seminal Plasma

Seminal plasma was collected from a vasectomized ram; analyses were done on 5 pools of 3 animals.

Electrophoresis

Polyacrylamide disc gel electrophoresis was carried out with 7.5 % running gel according to *Davis* (3). The gels were dyed with amido schwartz; densitometric tracings were taken at 592 nm.

Enzymes

The different fluids in the system used (1) were diluted with buffer solutions and incubated 4 h with chromogenic substrates at 37 °C on a special filter paper which was inserted between two transparent plastic slides. The substrates and the pH of the buffers are shown in table I.

Free Amino Acids

Analyses were performed on an aminolyzer Optica by ion exchange chromatography (14). Samples were lyophylized with exact, known quantities of Norleucine and α-amino-β-guanidinopropionic acid as internal references. Protein precipitation was accomplished by the addition of 3 % sulfosalycylic acid (6). The mean standard error and relative percentage (mol/100 mol) of each amino acid are shown in table II.

Results

Proteins

Female Tract Secretions and Blood Plasma (fig. 1)

Samples of tubal and uterine secretions and blood plasma give similar protein patterns. Albumin is, by far, always the most important component of all the samples, but transferrin and the main groups of globulins are also present in female genital tract secretions.

Seminal Plasma and Blood Plasma (fig. 2)

Seminal plasma is characterized by the presence of acid proteins in the prealbumin zone. While albumin is also an important compound of seminal

Table I. Enzymatic activities observed in ovine tubal and uterine secretions, blood plasma and seminal plasma

Enzyme	Substrate	pH	Tubal secretions	Uterine secretions	Blood plasma	Seminal plasma
Alkaline phosphate	phenol phthalein diphosphate	7.5	+++	+++	+	+
Nonspecific phosphatase	β-naphthyl phosphate	7.5	+++	+++	++	+
Esterase (acetate)	β-naphthyl acetate	7.5	+++	+++	++	++
Esterase (butyrate)	β-naphthyl butyrate	7.5	++	++	++	++
Lipase (nonanoate)	β-naphthyl nonanoate	7.5	+	+	+	++
Aminopeptidase (alanyl)	*DL*-alanyl-β-naphthylamide	7.5	++	+	+	+
Aminopeptidase (leucyl)	*L*-leucyl-β-naphthylamide	7.5	+	+	+	+
Aminopeptidase (valyl)	valyl-β-naphthylamide	7.5	0	0	0	±
Trypsin	N-benzoyl-*DL*-arginine-β-naphthalamide	7.5	0	0	0	0
Chymotrypsin	N-benzoyl-*DL*-phenyl alanine-β-naphthylamide	7.5	+	+	+	+
Galactosidase	ONPG	7.5	0	0	0	0
Acid phosphatase	phenol phthalein diphosphate	5.4	+	+	+	++
Glucuronidase	naphthol AS glucuronide	5.4	0	+	0	±
Glucosaminidase	naphthol ASBI N-acetyl glucosaminide	5.4	0	±	0	0
β-Glucosidase	6-bromo-2-naphthol β-glucopyranoside	5.4	0	±	0	0

0 = Not detected; ± = doubtful activity. +, ++, +++ represent respectively weak, medium, and high activity.

Table II. Content of ninhydrin-positive compounds of ewe blood plasma, tubal and uterine fluids at estrus and of vasectomized ram seminal plasma

	Amino acids (μmol/100 ml) in										
	tubal secretions			uterine secretions			blood plasma			seminal plasma	
	mean	SE	%	mean	SE	%	mean	SE	%	mean	%
Taurine	8.0	0.9	1.9	11.3	1.1	3.0	3.5	0.6	1.1	3.7	0.7
Urea	+			+			+			+	
Asp. + Asp. NH_2	9.7	1.4	2.4	21.5	2.9	5.7	6.9	0.5	2.2	28.7	5.7
Threonine	+			+			14.6	0.9	4.7	17.0	3.4
Serine	1.7	0.3	0.4	4.0	0.7	1.1	10.0	0.8	3.2	22.0	4.4
Glu. + Glu. NH_2	43.0	4.1	10.5	41.2	1.6	10.9	40.5	3.7	13.1	34.5	6.8
Proline	+	4.1		+			+			+	
Glycine	141.9	14.4	34.5	68.9	5.9	18.3	55.7	3.8	18.0	112.8	22.3
Alanine	45.8	2.0	11.1	38.9	4.0	10.3	25.1	2.6	8.1	34.0	6.7
Citrulline	6.9	0.7	1.7	11.2	0.7	3.0	12.1	3.1	3.9	trace	
Ac.-α-aminobutyr.	trace			trace			0			trace	
Cystine	0			trace			trace			trace	
Valine	27.4	2.8	6.7	40.9	4.1	10.9	34.3	2.5	11.1	41.3	8.2
Methionine	6.3	0.6	1.5	4.5	0.8	1.2	2.6	0.3	0.8	3.6	0.7
Isoleucine	10.9	1.2	2.6	12.7	0.7	3.4	10.9	0.5	3.5	13.0	2.6
Leucine	24.5	1.9	6.0	28.4	1.8	7.5	20.3	0.9	6.6	33.7	6.7
Tyrosine	13.1	1.3	3.2	10.2	0.8	2.7	7.6	0.5	2.5	12.9	2.6
Phenylalanine	13.9	1.2	3.4	10.3	0.7	2.7	6.8	0.5	2.2	10.4	2.1
Ornithine	8.0	0.8	1.9	10.6	0.9	2.8	9.2	1.2	3.0	5.8	1.1
Lysine	29.7	2.3	7.2	28.7	1.3	7.6	19.7	1.8	6.4	69.5	13.8
1-Me.-histidine	trace			trace			trace			trace	
Histidine	10.0	1.0	2.4	10.2	0.9	2.7	7.9	0.8	2.5	24.2	4.8
NH_3	trace			trace			trace			trace	
Tryptophane	trace			trace			trace			trace	
Arginine	13.5	0.9	3.3	22.9	1.3	6.1	22.4	1.9	7.2	37.0	7.3
Total	411.3			376.4			310.1			504.1	

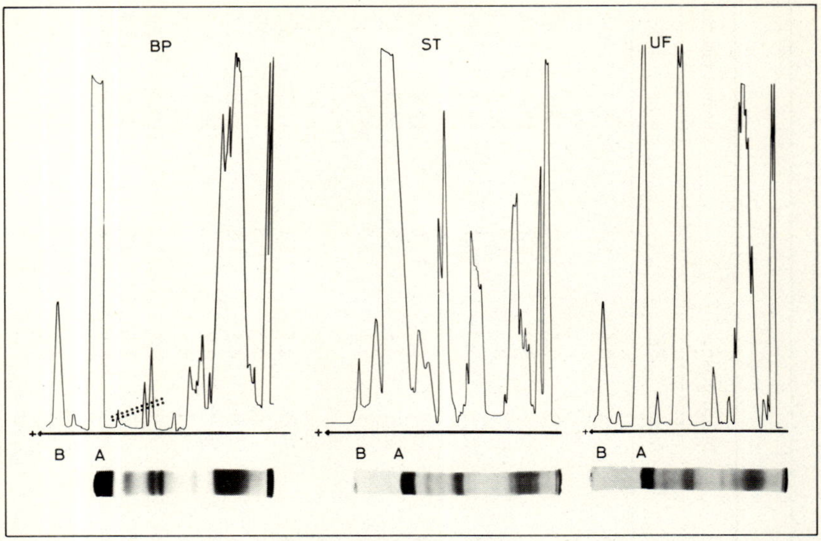

Fig. 1. Disc gel electrophoresis of blood plasma (BP), ewe tubal fluid (ST), and uterine fluid (UF). A Albumin; B bromophenol blue.

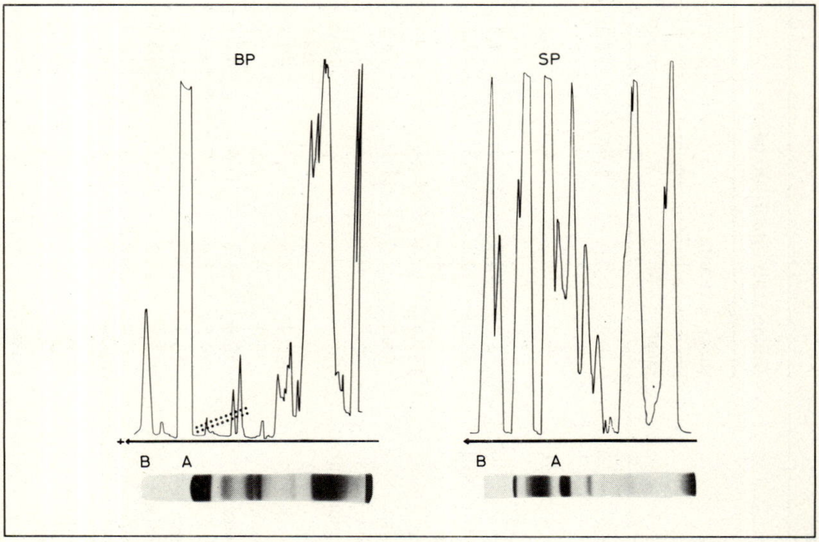

Fig. 2. Disc gel electrophoresis of blood plasma (BP) and vasectomized ram seminal plasma (SP). A Albumin; B bromophenol blue.

plasma, the other globulins having basic or high molecular weight are much less important than the serum.

Enzymes (table I)

Female Genital Secretions and Blood Plasma

Nonspecific esterases and phosphatases were detected in all the samples. The substrates of the amino peptidases are hydrolyzed, except for those of trypsic activity and only one amino peptidase.

Seminal Plasma and Blood Plasma

Enzymatic patterns are not very different. However, acid phosphatase activity is higher in seminal plasma. A weak β-glucuronidase activity can also be detected in the fluid.

Free Amino Acids (table II)

Female Genital Tract Secretions and Blood Plasma

Four principal observations are clear: (1) At the time of estrous cycle, the free amino acid levels of tubal and uterine fluids are similar to those of blood plasma. (2) Free glycine is the major amino acid, but is found in higher percentage in tubal fluids. (3) All the intermediate compounds of the arginine cycle are present in the three types of fluids. (4) Sulphured amino acids are weakly represented.

However, while concentrations of threonine and serine are higher in blood plasma, the contrary is true for taurine. Finally, the level of aspartic acid + asparagine is greater in uterine fluids, but arginine is significiantly lower in tubal fluids.

Seminal Plasma and Blood Plasma

If the total concentration of free amino acids is higher in seminal plasma, it is essentially due to aspartic acid + arginine, glycine, serine and basic amino acids, especially lysine.

Discussion and Conclusion

Tubal and uterine fluids seem to arise mainly from a transudation of plasma compounds. Seminal plasma seems to be less dependent on the vascular compartment and male genital tract, even if testicular and epididymal secretions are not concerned; it also has a high synthesizing activity.

Proteins

The protein composition of the medium bathing the spermatozoa differs during migration in the female genital tract when compared with seminal plasma. A dilution of these acid proteins (like the acrosin inhibitor and the decapacitation factor) occurs along with an increase in the relative amount of albumin.

Enzymes

In the ewe as in the rabbit (5), the major part of the enzymes described in the tubal epithelium of different species (9) can be detected in the fluid secreted. It is possible that these enzymes come from a filtration of blood plasma. However, with the technique employed the contribution of leukocytes to tubal and uterine zymograms cannot be determined, and there are only a few of them in the estrous cycle. The enzymes described in seminal plasma have already been observed in the ram and boar (4, 15). Seminal, tubal and uterine enzymatic patterns are qualitatively similar for most of the enzymes studied. The few differences observed may be due to difference in enzyme concentration or activity. However, female genital tract esterases and phosphatases can interfere in the nutritional metabolism of sperm (18), and aminopeptidases in the fertilization process (16) or the decapacitation and elimination of dead spermatozoa (7).

Free Amino Acids

As has already been observed (12), there is very high glycine concentration in the oviduct. The higher levels of some amino acids in blood plasma and in female genital tract secretions (glycine, threonine, serine, taurine) suggest that the uterine and oviductal walls are not passive filters. Our data on seminal plasma are similar to those of *Setchell et al.* (13) with high levels of glycine, alanine and lysine. Relations in the vascular compartment are also complex for the male tract. The role of free amino acids in sperm nutrition is not known, but it appears that during sperm ascent, the female genital tract provides a balanced, complete environment similar to that which the spermatozoa inhabited before ejaculation.

In the data presented, it appears that, except for nonenzymatic proteins, spermatozoa do not undergo a very great change in medium when carried in tubal or uterine fluids.

References

1 *Buissière, J.; Fourcard, A. et Colobert, L.:* Usage de substrats synthétiques pour l'étude de l'équipement enzymatique de microorganismes. C.R. Acad. Sci. *265:* 979–983 (1967).

2 *Chang, M.C.:* Detrimental effect of seminal plasma on the fertilizing capacity of sperm. Nature, Lond. *179:* 258–259 (1957).

3 *Davis, B.J.:* Disc gel electrophoresis. II. Method and application to human serum proteins. Ann. N.Y. Acad. Sci. *121:* 404–427 (1964).

4 *Dostal, J. and Veselsky, L.:* Proteins in the seminal plasma and the accessory sexual fluids of the bear. J. Reprod. Fertil. *30:* 255–267 (1972).

5 *Fredricsson, B.:* in *Hafez and Blandau* The mammalian oviduct, p. 326 (University of Chicago Press, Chicago 1969).

6 *Hamilton, P.B.:* Ion exchange chromatography of aminoacids. Microdetermination of free amino acids in serum. Ann. N.Y. Acad. Sci. *102:* 55–75 (1962).

7 *Joshi, M.S.; Yaron, A., and Lindner, M.R.:* An endopeptidase in the uterine secretion of the proestrus rat and its relation to a sperm decapitating factor. Biochem. biophys. Res. Commun. *38:* 52–57 (1970).

8 *Mattner, P.E.:* Capacitation of ram spermatozoa and penetration of the ovine egg. Nature, Lond. *199:* 772–773 (1963).

9 *Menezo, Y. et Laviolette, P.:* Les constituants aminés des sécrétions tubaires chez la lapine. Ann. Biol. anim. *12:* 383–396 (1972).

10 *Miyamoto. H. and Chang, M.C.:* The importance of serum albumin and metabolic intermediates for capacitation of spermatozoa and fertilization of mouse eggs *in vitro.* J. Reprod. Fertil. *32:* 193–205 (1973).

11 *Perkins, J.L.; Goode, L.; Wilder, W.A., and Henson, D.B.:* Collection of secretions from the oviduct and uterus of the ewe. J. anim. Sci. *24:* 383–387 (1965).

12 *Perkins, J.L. and Goode, L.:* Free amino acids in the oviduct fluid of the ewe. J. Reprod. Fertil. *14:* 309–311 (1967).

13 *Setchell, B.P.; Hinks, N.T.; Voglmayr, J.K., and Scott, T.W.:* Aminoacids in ram testicular fluid and semen and their metabolism by spermatozoa. Biochem. J. *105:* 1061–1066 (1967).

14 *Spackman, D.H.; Stein, W.H., and Moore, S.:* Automatic recording apparatus for use on the chromatography of aminoacids. Analyt. Chem. *32:* 1190–1206 (1958).

15 *Suska-Brezezinska, E.:* The presence of two aminopeptidases in the ram seminal fluid. Bull. Acad. Pol. Sci. *18:* 523–525 (1970).

16 *Toyoda, Y. and Chang, M.C.:* Sperm penetration of the rat egg *in vitro* after dissolution of zona pellucida by chymotrypsin. Nature, Lond. *220:* 589–590 (1969).

17 *Weinman, D.E. and Williams, W.L.:* Mechanism of capacitation of rabbit spermatozoa. Nature, Lond. *203:* 423–424 (1964).

18 *White, I.G. and Wallace, J.C.:* Breakdown of seminal glyceryl phosphorylcholine by secretions of the female reproductive tract. Nature, Lond. *189:* 843–844 (1961).

Author's address: *Y. Menezo,* Institut National de la Recherche Agronomique, Laboratoire de Biologie 406, INSA, *F–69621 Villeurbanne* (France)

The Biology of Spermatozoa
INSERM Int. Symp., Nouzilly 1973, pp. 182–191 (Karger, Basel 1975)

The Need for Capacitation of Human Sperm: Functional and Ultrastructural Observations

Pierre Soupart

Department of Obstetrics and Gynecology,
Vanderbilt University School of Medicine, Nashville, Tenn.

The study of human sperm capacitation requires the use of an *in vitro* fertilization system, the effectiveness of which has to be demonstrated prior to discussing functional and ultrastructural evidence for human sperm capacitation. The problem of authenticating fertilization of human oocytes in culture was complicated by the failure of conventional light microscopic methods to demonstrate the presence of the fertilizing sperm flagellum within the ooplasm of oocytes at late pronuclear stages (24 h or more postinsemination, fig. 1). Figure 1 a–c shows two of the fertilization criteria (2 pronuclei, 2 polar bodies) in such a specimen, as well as the lack of the third and critical criterion, the flagellum, after fixation and staining (fig. 1 d). Such an observation raised the very fundamental question of whether such a specimen, despite its appearance, was authentically fertilized or the result of artificial activation, mimicking exactly the configuration of the fertilized pronuclear ovum (1). This prompted an intensive ultrastructural study of fertilized human ova at late pronuclear stages (2). It was found that the failure of conventional methods to demonstrate the presence of the fertilizing sperm flagellum was due to the rapid disappearance of the most heavily staining structural elements of the sperm tail (fig. 2). Even high resolution light microscopy was unable to demonstrate the sperm flagellum, although it was effectively present as revealed by electron microscopy. At late pronuclear stages, the only structural element of the flagellum to remain identifiable in the ooplasm, is the axial filament complex (fig. 2 d, 3). In favorable sections (fig. 3), all fertilization criteria can be identified at once. A small lobe or fragment of the first polar body (PB_1), characterized by numerous electron-dense cortical granules, is visible in the perivitelline space. The second polar body (PB_2) is almost totally deprived of cortical granules. Cortical granules are also practically absent from underneath the vitelline membrane. The female pronucleus (FP) is identified by its topographical relationship with the second

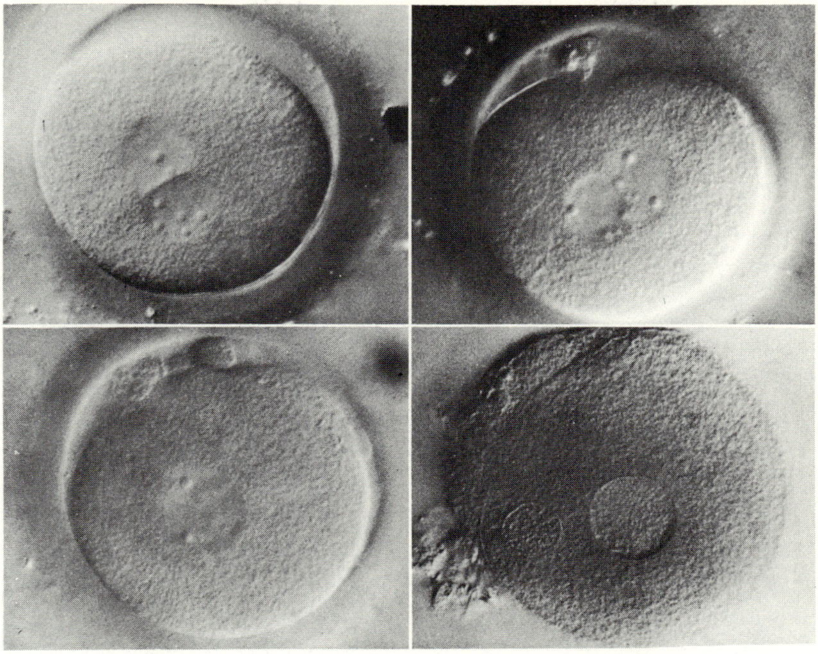

a, b

c, d

Fig. 1. Failure of a conventional method (fixation: acetic acid-methanol; stain: aceto-carmine) to demonstrate remnants of the fertilizing sperm flagellum in the ooplasm of a human ovum at late pronuclear stage (see text). Nomarski. × 350.

polar body, and the male pronucleus (MP) is identified by the presence close to its edge of the axial filament complex (AFC, inset), which was the only structural element of the flagellum to remain identifiable in specimens fixed 24 h postinsemination. Such late pronuclear stages appeared to be reached in culture within physiological time limits (3).

The problem of mammalian sperm capacitation has been exhaustively reviewed in recent years by others and by ourselves (4). In order to recognize the need for capacitation of human sperm, two types of evidence are necessary. First, sperm ability to penetrate through the zona pellucida should be shown to be significantly increased as a result of a given treatment (functional evidence). Second, sperm traversing the zona pellucida should be shown to have undergone the acrosome reaction (ultrastructural evidence). In their 1969 report, *Edwards et al.* (5) suggested that capacitation had resulted spontaneously from environmental conditions created by Bavister's medium. Later (6), it was suggested that the acrosome reaction could have resulted from the action of progesterone produced by follicle cells of oocytes having completed much of their maturation *in*

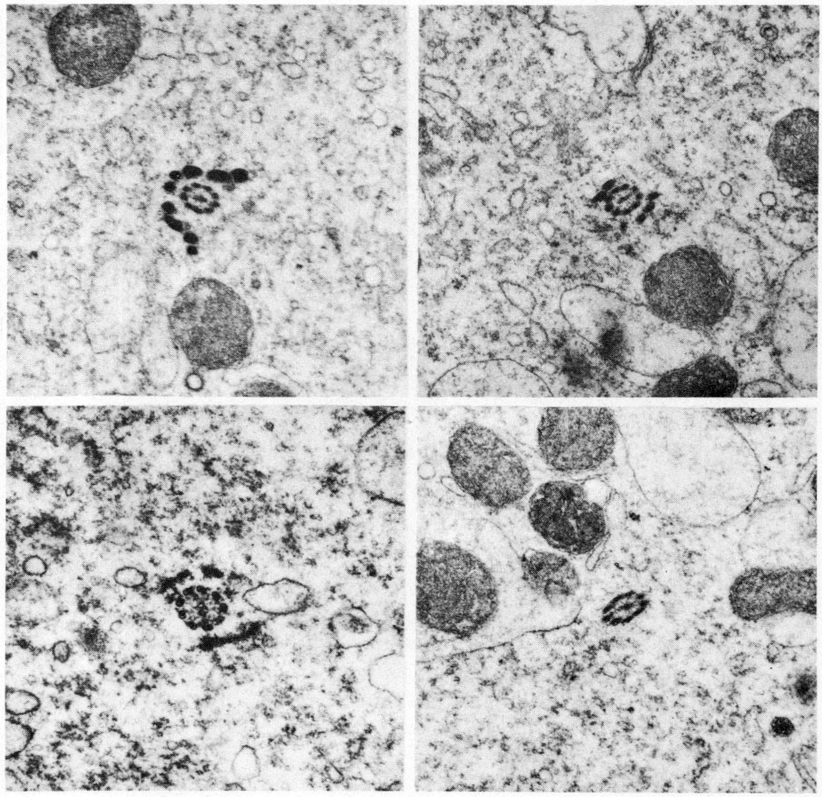

a, b

c, d

Fig. 2. Progressive dismantling of the fertilizing sperm flagellum within the ooplasm. *a,* Opening of the 9 outer dense fibers of the midpiece (× 22,100). *b,* Four of the dense fibers have disappeared (× 16,750). *c,* Breakdown of the hemicircumferential ribs of the principal piece (× 32,400). *d,* The axial filament complex is the last structure to remain identifiable (× 20,100).

vivo following stimulation by exogenous gonadotropins. Since in these reports the earliest time at which sperm were observed in the perivitelline space was 7.5 h postinsemination, it seemed more likely that some cellular interaction was needed to bring about capacitation, rather than simple environmental conditions determined by the fertilization medium. In 1971, *Soupart and Morgenstern* (7) designed a series of experiments in order to determine the effects of gonadotropin (FSH, LH and HCG) stimulation of follicle cells on the human sperm ability to penetrate the zona pellucida *in vitro.* The results of these experiments clearly indicated that the presence of exogenous gonadotropins in the fertilization environment increased sperm penetrating ability in a statistically significant manner. Penetration to various degrees was observed only in oocytes initially surrounded by follicle cells. All oocytes that were naked at the start of fertilization cultures were found later on with sperm attached to the outer surface of their zona, but no degree of penetration had been achieved at the end of individual culture periods. It was concluded from these initial experiments that the hormonally induced enhancement of sperm penetrating ability was the first functional evidence of the need for capacitation of human sperm, and that the observed effect was most likely mediated by gonadotropin-stimulated corona cells. The morphological evidence of human sperm capacitation was obtained from the ultrastructural study of oocytes matured and fertilized *in vitro*. Maturation was obtained in culture using a two-stage system, in which immature oocytes were first primed with 17β-estradiol (0.1 μg/ml), then cultured for 40–44 h in the presence of 17α-hydroxyprogesterone (0.1 μg/ml), the basic medium being Ham's F10 in both cases. Fertilization cultures were carried out

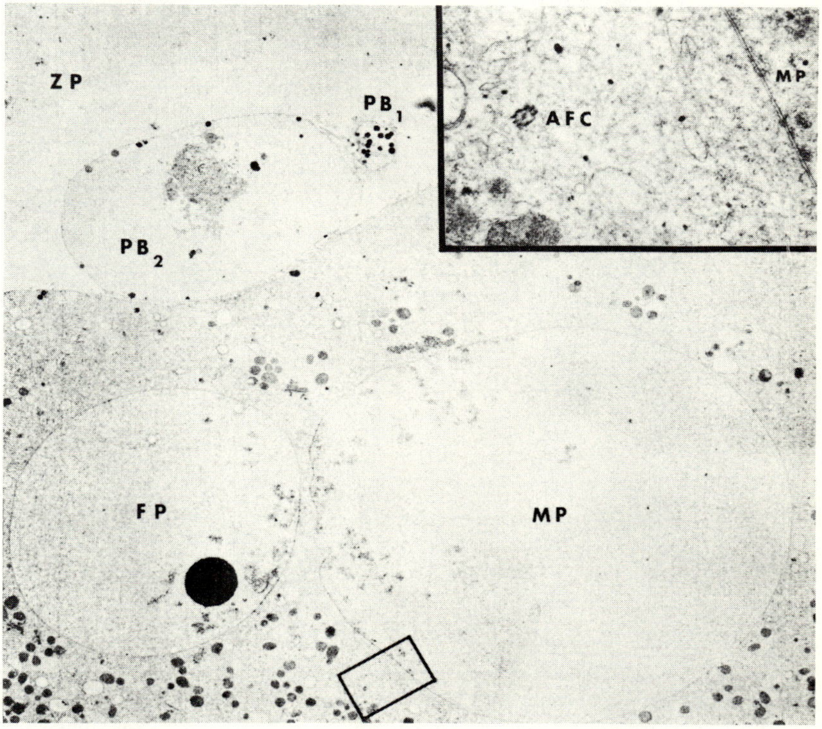

Fig. 3. Criteria of fertilization (see text) in a human pronuclear ovum fixed 24 h postinsemination (× 4,400). Rectangular area indicates the location of a cross section through the axial filament complex, shown at higher magnification in the inset. × 25,900.

using the system previously described (7). Under such conditions, 44.2 % of 154 oocytes did mature in culture, about one-fifth of which underwent normal monospermic fertilization (2). In such specimens, two types of sperm were consistently observed in close vicinity or direct contact to the zona pellucida (fig. 4). The first type (fig. 4a) presented all the morphological characteristics of the most normal-appearing spermatozoa described to be present in the human ejaculate (8–12). Since the specimens were fixed 24 h or more postinsemination, it appears that sperm do not necessarily undergo membrane breakdown as a result of being washed, centrifuged, diluted to appropriate concentration, and

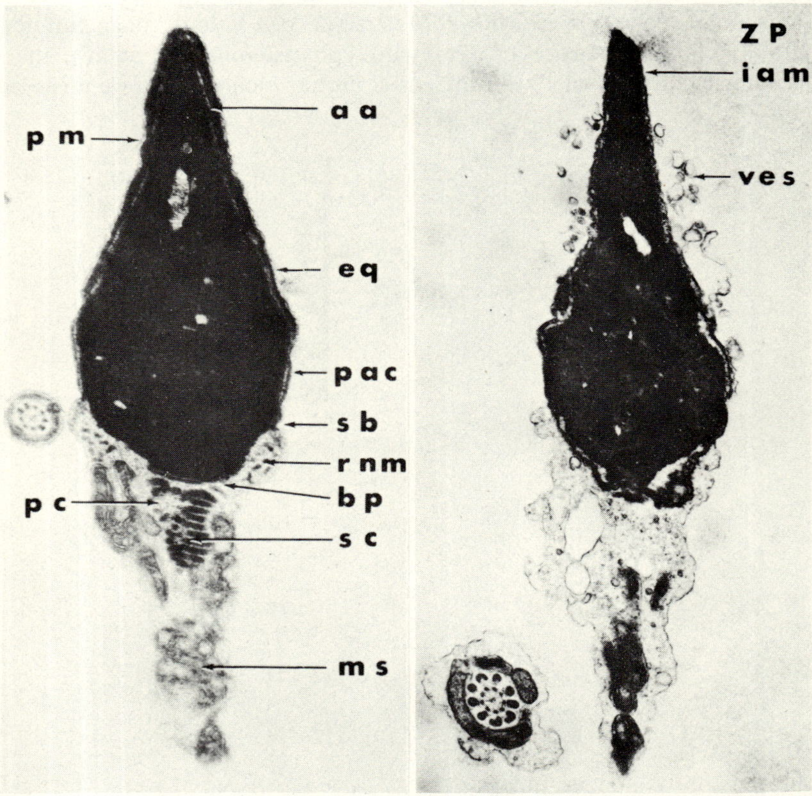

a, b

Fig. 4. The two types of sperm currently observed in close vicinity to the zona pellucida. *a,* Intact sperm; *b,* sperm with reacted acrosome. pm, Plasma membrane; aa, anterior acrosome; eq, equatorial segment; pac, postacrosomal cap; sb, striated band; rnm, redundant nuclear membrane; bp, basal plate; pc, proximal centriole; sc, striated column; ms, mitochondrial sheath; ZP, zona pellucida; iam, inner acrosomal membrane; ves, vesiculation products. × 22,900.

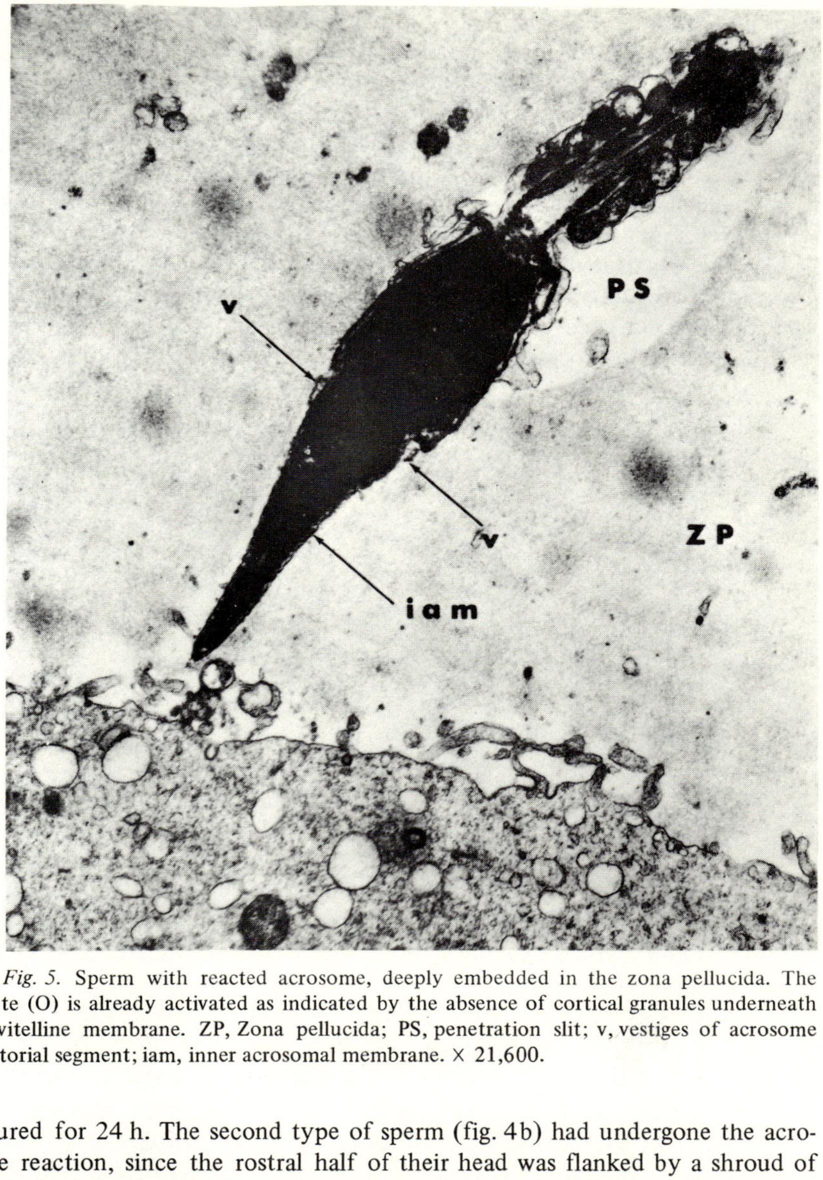

Fig. 5. Sperm with reacted acrosome, deeply embedded in the zona pellucida. The oocyte (O) is already activated as indicated by the absence of cortical granules underneath the vitelline membrane. ZP, Zona pellucida; PS, penetration slit; v, vestiges of acrosome equatorial segment; iam, inner acrosomal membrane. × 21,600.

cultured for 24 h. The second type of sperm (fig. 4b) had undergone the acrosome reaction, since the rostral half of their head was flanked by a shroud of acrosomal vesiculation products. In sperm observed within the zona pellucida, the inner acrosomal membrane was in direct contact with the zona material (fig. 5). The sharpness of the penetration slit and the even electron density of its edges suggest that the acrosomal proteinase (13, 14) is firmly bound to the inner acrosomal membrane. Thus, clearly, the human spermatozoon passing through the zona pellucida of the human oocyte had undergone the acrosome reaction. In the light of current concept of capacitation and of the present observations, both functional and ultrastructural, it follows that the human sperm had to be capacitated first. On specimens fixed 24 h postinsemination, the majority of sperm seen outside the zona had undergone the acrosome reaction and were deeply embedded in between corona cells, or sequestered by corona cells (2). No vesiculation products or strip of membranes were to be seen close to the sperm heads, which indicated that such products were left behind during sperm progression in between corona cells. These morphologic observations are in keeping with the concept, derived from functional observations (7), that the human sperm capacitation process is mediated by gonadotropin-stimulated follicle cells.

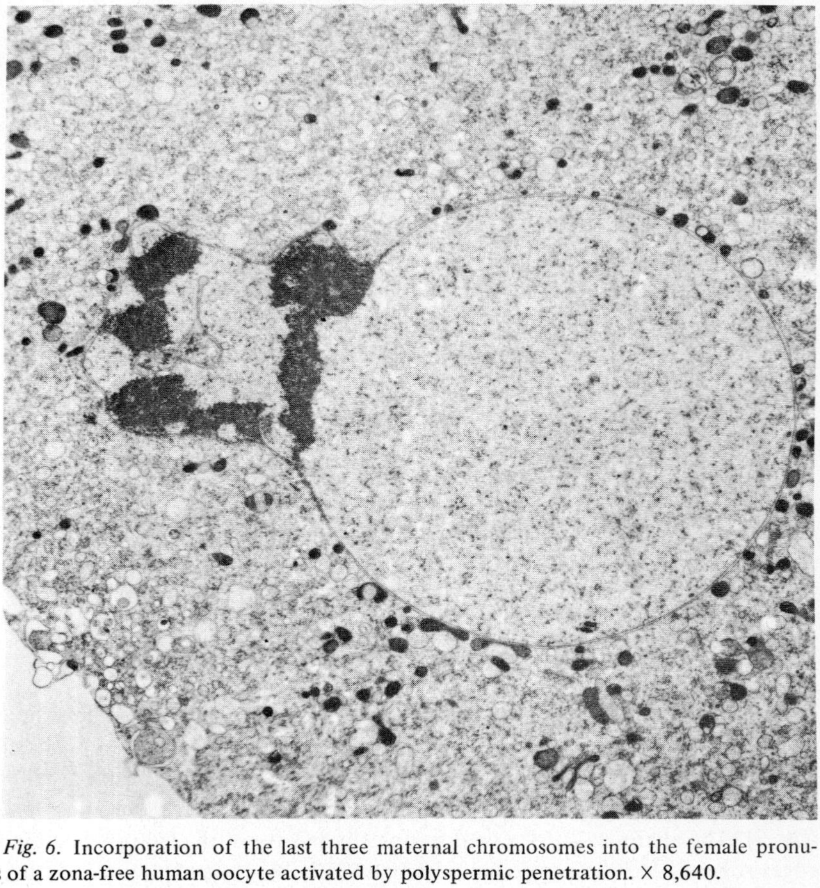

Fig. 6. Incorporation of the last three maternal chromosomes into the female pronucleus of a zona-free human oocyte activated by polyspermic penetration. × 8,640.

a, b

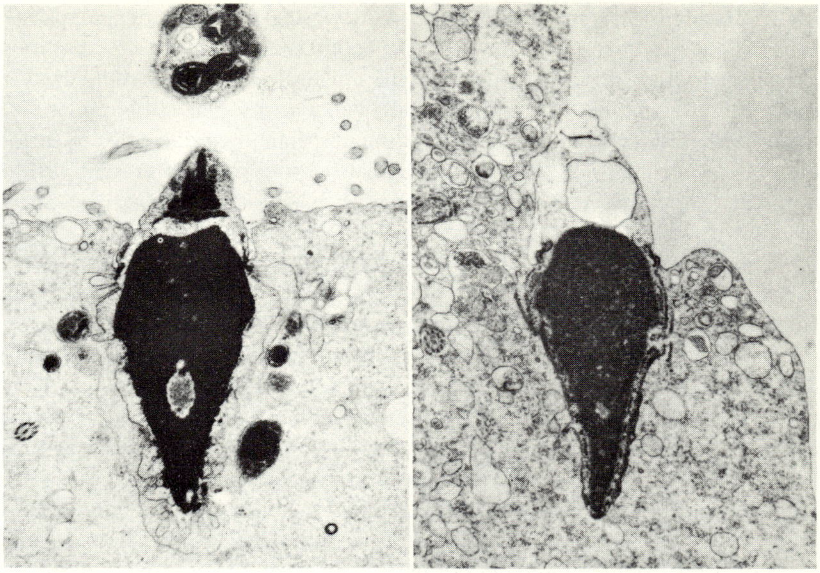

Fig. 7. Fusion of sperm with reacted (*a:* × 15,400) and unreacted (*b:* × 16,200) acrosome with zona pellucida-free human oocytes.

If capacitation consists essentially of the initial conditioning which will ultimately confer to the sperm their ability to pass through the zona pellucida, it has also been suggested that capacitation may involve changes in the postnuclear cap which determine sperm ulterior ability to fuse with the oocyte (15). To further investigate this possibility, inseminated zona pellucida-free oocytes were studied with the aid of the electron microscope (16). All specimens underwent polyspermic fertilization, from moderate to extreme degree. Various stages of male pronucleus development were observed which are described elsewhere (16). The zona-free oocytes had matured in culture and became activated upon insemination, as shown by the absence of cortical granules and the development of a female pronucleus (fig. 6). Considering that supplementary sperm were never observed in the perivitelline space of normally fertilized ova with intact zona pellucida, the ineluctable conclusion to be drawn is that the zona is the essential if not unique site of the block to polyspermy. The zona is therefore of utmost functional importance in the human fertilization process since polyploidy is a lethal condition in man. Furthermore, the study of activated zona-free human oocytes provided information of the early stages of gamete fusion. In figure 7, two spermatozoa are shown at the same stage of fusion with the oocyte, as evidenced in both cases by the continuity of the vitelline membrane and the sperm plasma membrane at the level of the posterior margin of the postnuclear cap. In one specimen (fig. 7a), the only membranous structures surrounding the sperm nucleus are the two leaflets of the nuclear membrane pulling away from the sperm chromatin. There is no evidence of acrosomal structure, indicating that the spermatozoon had undergone the acrosome reaction prior to fusion. In the other specimen (fig. 7b), all acrosomal features are still present within the ooplasm, indicating that this spermatozoon had not undergone the acrosome reaction prior to fusion. It appears that, at least under the particular conditions of these observations, the lack of acrosome reaction and presumably therefore the lack of capacitation, did not interfere with the fusion of human gametes when the zona pellucida did no longer prevent their immediate contact.

In the light of these observations, there can be no longer any doubt as to the need of capacitation of human sperm as a prerequisite for fertilization. Although much work remains to be done to disclose the molecular mechanisms of capacitation and acrosome reaction, the limiting factor of human *in vitro* fertilization is no longer the problem of obtaining capacitated sperm, but the problem of mastering culture conditions leading to complete control over oocyte maturation *in vitro.* If the induction of ovulation by current techniques (HMG-HCG) provides oocytes having completed much of their maturation *in vivo,* it throws no light on the factors regulating maturation, and that is the next challenge facing the student of human fertilization.

References

1 *Austin, C.R.:* The mammalian egg, p. 27 (Blackwell, Oxford 1961).

2 *Soupart, P. and Strong, P.A.:* Ultrastructural observations on human oocytes fertilized *in vitro.* Fertil. Steril. *25:* 11–44 (1974).

3 *Edwards, R.G.:* Fertilization and cleavage *in vitro* of human ova; in *Moghissi and Hafez* Biology of mammalian fertilization and implantation, pp. 263–278 (Thomas, Springfield 1972).

4 *Soupart, P.:* Sperm capacitation: methodology, hormonal control and the search for a mechanism; in *Moghissi and Hafez* Biology of mammalian fertilization and implantation, pp. 54–125 (Thomas, Springfield 1972).

5 *Edwards, R.G.; Bavister, B.D., and Steptoe, P.C.:* Early stages of fertilization *in vitro* of human oocytes matured *in vitro.* Nature, Lond. *221:* 632 (1969).

6 *Edwards, R.G.; Steptoe, P.C., and Purdy, J.M.:* Fertilization and cleavage *in vitro* of preovulator human oocytes. Nature, Lond. *227:* 1307 (1970).

7 *Soupart, P. and Morgenstern, L.L.:* Human sperm capacitation and *in vitro* fertilization. Fertil. Steril. *24:* 462 (1973).

8 *Pedersen, H.:* Ultrastructure of the ejaculated human sperm. Z. Zellforsch. *94:* 542 (1969).

9 *Zamboni, L.; Zemjanis, R., and Stefanini, M.:* The fine structure of monkey and human spermatozoa. Anat. Rec. *169:* 129 (1971).

10 *Zamboni, L. and Stefanini, M.:* The fine structure of the neck of mammalian spermatozoa. Anat. Rec. *169:* 155 (1971).

11 *Pedersen, H.:* The post-acrosomal region of spermatozoa of man and *Macaca arctoides.* J. Ultrastruct. Res. *40:* 366 (1972).

12 *Stefanini, M.; Martino, C. de, and Zamboni, L.:* Fixation of ejaculated spermatozoa for electron microscopy. Nature, Lond. *216:* 173 (1967).

13 *Stambaugh, R. and Buckley, J.:* Zona pellucida dissolution enzymes of the rabbit sperm head. Science *161:* 585 (1968).

14 *Stambaugh, R. and Buckley, J.:* Identification and subcellular localization of the enzymes effecting penetration of the zona pellucida by rabbit sperm. J. Reprod. Fertil. *19:* 423 (1969).

15 *Yanagimachi, R. and Noda, Y.D.:* Ultrastructural changes in the hamster sperm head during fertilization. J. Ultrastruct. Res. *31:* 465 (1970).

16 *Soupart, P. and Strong, P.A.:* Ultrastructural observations on polyspermic penetration of zona pellucida-free human oocytes inseminated *in vitro* Fertil. Steril. (in press, 1975).

Author's address: Prof. *P. Soupart,* Department of Obstetrics and Gynecology, School of Medicine, Vanderbilt University, *Nashville, TN 37232* (USA)

The Biology of Spermatozoa
INSERM Int. Symp., Nouzilly 1973, pp. 192–200 (Karger, Basel 1975)

The Acrosomal Enzymes of Mammalian Spermatozoa

Lourens J.D. Zaneveld

Population Research Center, Medical Sciences and Engineering Division, IIT Research Institute, Chicago, Ill.

Introduction

Spermatozoa possess a multitude of enzymes and enzyme systems. These can be divided into two categories: those involved with sperm metabolism (associated with the midpiece), and those with other functions (mostly associated with the sperm head). The enzymes of the second category are similar to the ones present in lysosomes and probably originate to a large extent from the acrosome, a structure that forms a bag-like cap on the anterior surface of the sperm head. It is produced by the Golgi apparatus during the process of spermatogenesis. Evidence for the location of an enzyme in the acrosome has to come from morphological, immunological or histochemical studies, however, since no chemical or physical treatment has the capability of producing absolutely uncontaminated acrosomal preparations. Only four enzymes were unquestionably shown to be located within the acrosome: aryl sulphatase (24), hyaluronidase (13), a nonspecific esterase (3) and a proteinase called acrosin (37). Acid phosphatase is present in the post-acrosomal cap and in certain parts of the equatorial segment (35). Other enzymes that may be associated with the acrosome, or at least with the sperm head, are arylamidase (15), alkaline phosphatase (1, 4), β-glucuronidase (4), ribonuclease (4), acid proteinase (1, 4, 22), β-N-acetylglucosaminidase (1), phospholipase A (1), neuraminidase and neuraminidase-like factor (29), and an enzyme that on addition to the ovum disperses the corona radiata ('corona-penetrating enzyme', CPE) (45). Many of these enzymes can also be found in the cytoplasmic droplets (4). Such droplets are remnants of the original spermatid cell (the last cell stage during sperm formation in the testis – spermatogenesis) that move downward towards the end of the midpiece while the spermatozoa are transported through the epididymis and are usually removed during ejaculation.

Some indications are present for the physiological function of four of these enzymes: hyaluronidase, CPE, acrosin and the neuraminidases. Their role is mostly related to the fertilization process, i.e. sperm penetration through the external layers of the ovum. These layers include the cumulus oophurus (outermost layer), corona radiata (middle layer) and the zona pellucida (innermost layer). The egg itself is surrounded by the vitelline membrane. Although sperm motility also plays a major role in the fertilization process, union of the gametes cannot take place without the action of the acrosomal enzymes since spermatozoa penetrate the ovum with its layers completely intact. Fairly good evidence is available that the lytic action of hyaluronidase allows the passage of the spermatozoon through the cumulus, that of CPE through the corona and that of acrosin through the zona. Sperm neuraminidase may either alone or together with cortical granule material induce the zona reaction (block to polyspermy) of the egg.

The following is a discussion of these enzymes in sequence as they appear to be used by the spermatozoon to penetrate the ovum. Biochemical details are omitted but these may be obtained from the referenced publications. Particular emphasis will be placed on the application of our knowledge concerning the acrosomal enzymes to the development of contraceptives, since the inhibition of these enzymes has been shown to prevent fertilization. For such purposes, two types of anti-enzymes may be used: enzyme inhibitors and enzyme antibodies. Enzyme inhibitors have the advantage that they are usually of low molecular weight and can therefore rapidly diffuse through tissues and membranes to contact the enzyme. They are usually not highly specific for one particular enzyme, however. Antibodies may also have enzyme-inhibiting properties although they are generally not as effective as the inhibitors. Because of their high molecular weight, it is presently questionable if systemic antibodies can enter the female genital tract and be present in sufficient concentrations at the site of fertilization. Local antibody production in the cervix may be the answer to this problem. The antibodies are normally very specific for their respective antigens, however, and both anti-enzymes deserve evaluation for their effectiveness as contraceptive agents.

Hyaluronidase

Hyaluronidase was the first acrosomal enzyme with a known function in the reproductive process. Large variations exist among species in regard to the hyaluronidase content of spermatozoa: the ram, bull and rabbit having high amounts and the rat, stallion and rooster possessing very low amounts (40). Ejaculated and epididymal spermatozoa possess approximately the same amount of hyaluronidase. As with the other acrosomal enzymes, spermatozoa do not actively produce hyaluronidase but liberate this enzyme under various conditions. All the hyaluronidase of semen originates from spermatozoa and a direct correlation exists between the amount of seminal hyaluronidase and the concentration of spermatozoa, but not their morphology or motility. A hormone-dependent increase in hyaluronidase release occurs from the spermatozoa after residence in the uterus, or on addition of uterine fluid *in vitro* (12). This increase in sperm hyaluronidase activity is not due to the removal of an inhibitor (as is the case with acrosin) since no such inhibitor is present in seminal plasma or has ever been found on spermatozoa. Interestingly, such hyaluronidase inhibitors are present in blood serum and may prevent the removal of the cumulus cells by sperm hyaluronidase (11). No relationship exists, however, between female sterility and the concentration of serum hyaluronidase inhibitors (11). The absence of hyaluronidase from semen renders the sperm infertile (10), whereas the addition of hyaluronidase to spermatozoa before deposition into the vagina may enhance their fertilizing capacity (10, 23). The latter has been questioned and could only be explained if the hyaluronidase adheres to the spermatozoa since seminal fluid or other solutions are normally not transported through the female genital tract. Although these observations should be checked in more detail, they do imply the essential role of hyaluronidase for successful fertilization.

Sperm hyaluronidase has been studied quite extensively and is identical to testicular hyaluronidase, although it appears to be different from the hyaluronidases of other tissues and organs (16, 40). Similar to acrosin, the enzyme may therefore be unique to the male reproductive tract. Hyaluronidase inhibitors that prevent the dispersion of the cumulus cells *in vitro,* also decrease the fertilizing capacity of spermatozoa either on direct addition to the sperm or if placed in the vagina before coitus (19, 20). Good hyaluronidase inhibitors of low molecular weight are as yet not available, possibly explaining why further research in this area has not been performed. While heterologous and isologous hyaluronidase antibodies were shown to inhibit the activity of the enzyme and prevent the dissolution of the follicle cells by spermatozoa or their adherence to the ovum (16, 17, 40), it has never been shown that the induction of specific hyaluronidase antibodies in a female animal interferes with her reproductive capacity. It was shown, however, that 12.5 % of all infertile women tested possessed hyaluronidase-inhibiting antibodies (16). Further evaluation of hyaluronidase inhibitors as antifertility agents appears necessary.

Corona-Penetrating Enzyme (CPE)

This enzyme was not discovered until recently (45). It can be obtained from spermatozoa by detergent extraction and is of particular interest since its activity is inhibited by the decapacitation factor (DF) (45). DF is present in seminal plasma in a high molecular weight form and is as such located on ejaculated spermatozoa. It was the first seminal component shown to prevent fertilization on addition to capacitated spermatozoa (36). The seminal trypsin inhibitors have a similar effect as will be discussed in the following section. DF can be reduced to a small peptide that still possesses active antifertility properties (36). Although DF is normally removed from the sperm during capacitation, it is certainly feasible that a slight biochemical modification of the active peptide may cause it to bind irreversibly to the sperm and CPE, and hence be an effective contraceptive agent. One advantage of this particular compound would be that it occurs naturally and may have less side effects than synthetic inhibitors. High molecular weight DF may have a stabilizing effect on the acrosome, but in its low molecular weight form, it probably acts strictly as a CPE inhibitor. This is further illustrated by the fact that DF treated, capacitated spermatozoa cannot fertilize cumulus free ova, unless the corona radiata has been removed (6). The antifertility effect of DF lies in its ability to prevent corona penetration by spermatozoa, therefore. Additionally, DF does not inhibit the activity of hyaluronidase or proteinases (45). Other CPE inhibitors that were shown to have antifertility properties are the glycoproteins fetuin and boar Cowper's gland mucin (28). Univalent and bivalent antirabbit sperm antibodies (produced against both ejaculated and capacitated spermatozoa) effectively prevent sperm attachment and passage through the corona (16), indicating that this enzyme is also susceptible to inhibition by antibodies.

Acrosin

The sperm neutral proteinase (Acrosin, IUB No. 3.4.21.10) has been studied quite extensively and may be treated as a model for other acrosomal enzymes for the development of new birth control methods. Interest in this enzyme developed when it was shown that the dispersion of the zona pellucida from ova by sperm extracts was due to a proteinase (30). Detailed studies showed that acrosin differs biochemically from all other known proteolytic enzymes although it has many properties in common with trypsin and plasmin (5, 21, 39, 42). Immunological differences exist between these last two enzymes and acrosin also (44), and this sperm proteinase appears specific to the male reproductive tract.

Proteinase inhibitors are present in the seminal plasma of all mammals. No correlation appears to exist between the amount of inhibitor and the motility or abnormal forms of spermatozoa, although a low concentration of sperm is usually associated with low amounts of inhibitor (9). The seminal inhibitors can be divided into two groups: those with a rather low molecular weight (5,000–15,000) and those with higher molecular weights. The latter are identical to the serum proteinase inhibitors α_1-antitrypsin and α_{1X}-antichymotrypsin (34). The other serum inhibitors – inter-α-trypsin inhibitors, α_2-macroglobulin, antithrombin III and the C_1-esterase inhibitor – are absent from seminal plasma at least in the case of the human (34). The α_1-antitrypsin and the α_{1X}-antichymotrypsin are also present in oviductal fluid (31) and cervical mucus (26) having their highest concentration at midcycle.

All mammalian spermatozoa studied so far possess acrosin. In various species, including the rabbit, boar and stallion, acrosin activity is high in epididymal spermatozoa but is much less in ejaculated sperm (43). This is caused by the addition of one or more of the seminal proteinase inhibitors to spermatozoa during ejaculation, further evidence of which is the similarity in molecular weight (MW) and enzyme-inhibition spectrum between one of the low MW seminal plasma inhibitors and an inhibitor purified from washed spermatozoa (39). In other species, high levels of inhibitor are already associated with the spermatozoa as they are stored in the epididymis (43). The epididymal inhibitor most likely originates from the rete testis (33). Its concentration in boar epididymal fluid is much less than in boar seminal plasma, explaining the difference in acrosin activity of epididymal and ejaculated sperm in this species. The ram does not show such differences (43) and indeed the epididymal plasma and seminal plasma of the ram possess approximately the same concentration of inhibitor (33). Evidence is also available that the acrosin of testicular sperm may be partially in a zymogen form (14). The inhibitor(s) adhere tightly to spermatozoa since they cannot be removed by washing. The location of acrosin and its inhibitor(s) within the acrosome has not been firmly established although results point to their association with the internal and external membranes, respectively.

Capacitated spermatozoa possess a much higher acrosin activity than ejaculated sperm, apparently due to the removal of the inhibitor(s) during uterine incubation. The mechanism by which this occurs is not understood. Such removal is essential for successful fertilization as could be shown by the contraceptive effect of adding the Kunitz pancreatic trypsin inhibitor (K-PTI), low MW inhibitors isolated from both rabbit and guinea pig seminal plasma, or the rete testis inhibitor to capacitated spermatozoa (32, 41). Similar results could not be obtained with α_1-antitrypsin or α_{1X}-antichymotrypsin (38) and it appears unlikely that the higher MW inhibitors play a role in the fertilizing process.

The unique biochemical properties of acrosin and the antifertility effect of the low MW inhibitors encouraged further study of proteinase inhibitors as contraceptive agents. It was essential, however, to find an inhibitor that complexes irreversibly with the acrosin so that it would not be removed during the capacitation process. As mentioned in the previous paragraph, natural inhibitors do not fall into this category. This was further emphasized by the lack of inhibition of fertilization on mixing K-PTI with ejaculated spermatozoa and addition of the sperm to the vagina or uterus (25). Additionally, such irreversible inhibitors should be of very low molecular weight enabling them to diffuse rapidly through sperm membranes. Preferably, they should also be specific for acrosin. The first two criteria, and to some extent the third, were met by the use of synthetic, active site directed reagents. Tosyl-lysine chloromethyl ketone (TLCK) and nitrophenyl *p*-guanidino benzoate (NPGB) irreversibly inhibit the *in vitro* activity of acrosin, in contrast to tosylamide phenylethyl chloromethyl ketone (TPCK) that has no effect on the enzyme (40, 42). TLCK but not TPCK, in quantities that do not effect sperm motility, completely inhibited fertilization of capacitated spermatozoa (41) indicating that the synthetic inhibitors function by specific inhibition of acrosin and not by some other mechanism. It was further shown that TLCK specifically blocks sperm penetration through the zona pellucida *in vitro* (6). Both TLCK and NPGB prevent fertilization if mixed with ejaculated spermatozoa before addition to the uterus. TLCK also effectively prevents conception on vaginal application before coitus and enhances the contraceptive activity of Delfen vaginal cream (41). Two other synthetic inhibitors, *p*-aminophenacyl bromide (APB) and *p*-guanidino-phenacyl bromide (GPB), cause a decrease in fertilization if applied vaginally, but were less active than TLCK (18). A mixture of inhibitors appears to have the greatest contraceptive effect, however (18). TLCK, GPB, *p*-amino benzamidine hydrochloride, benzamidine hydrochloride, and to a lesser extent, 1,4-diamidinobenzene dihydrochloride also prevent the fertilization of frog eggs (8).

The activity of acrosin on high MW (gelatin) but not on low MW (BAEE) substrates can be inhibited by addition of acrosin antibodies (44). No antifertility experiments with such antibodies have been performed as yet and their contraceptive ability remains to be established.

Neuraminidases

Two types of neuraminidases were shown to be associated with spermatozoa: a neuraminidase identical to that present in other mammalian tissues, and a neuraminidase-like factor (NLF) that has a somewhat different substrate specificity (29). One or both of these neuraminidases may be involved in the zona reaction of the egg since the addition of partially purified neuraminidase preparations to unfertilized ova renders the zona as resistant to certain chemicals as if the egg had been fertilized (7). Additionally, the pretreatment of ova with neuraminidase greatly decreases both *in vivo* and *in vitro* fertilization by inhibition of sperm penetration through the zona (6, 27). The hypothesis would be that as the first sperm passes through the zona, the neuraminidase(s) is (are) released, triggering a reaction in the zona pellucida that renders it impenetrable to other spermatozoa. Other investigators have evidence, however, that the cortical granule material may have a similar effect on the zona (2). Their suggestion is that as the fertilizing spermatozoon enters the peri-vitelline space and induces the cortical reaction, the released cortical material causes the zona changes. It is of course possible that both sperm and cortical granule enzymes are involved in the zona reaction but this remains to be established. Since the sperm neuraminidases seem to prevent fertilization, their inhibition will not lead to infertility and they are therefore probably of no use in the development of antifertility agents.

Conclusion

At present only minimal knowledge is available concerning the physiological role of sperm acrosomal enzymes in the reproductive process. One thing for certain, however, is that these enzymes are essential for union of the gametes to occur. The idea of using sperm acrosomal enzyme inhibitors as contraceptive agents is by no means new. Enhanced interest in this area, coupled with the great need for an effective, nonhormonal birth control method, has led to various original concepts and experiments showing the practical application of such antienzymes. As yet, completely satisfactory compounds have not been found. Even TLCK that would appear promising may only be useful for vaginal application since it has some undesirable properties and inhibits trypsin, plasmin and thrombin as well as acrosin. Present results encourage further elucidation of such compounds, however, and the mechanisms by which they prevent fertility.

Acknowledgments

The unpublished results reported in this communication were supported by NIH Grant No. NIH-HD-06315, Ford Foundation Grant No. 690-0108 and The Biomedical Center for Population Research Grant No. NIH-HD-07110.

References

1 *Allison, A.C. and Hartree, E.F.:* Lysosomal enzymes in the acrosome and their possible role in fertilization. J. Reprod. Fertil. *21:* 501–515 (1970).

2 *Barros, C. and Yanagimachi, R.:* Induction of the zona reaction in golden hamster eggs by cortical granule material. Nature, Lond. *233:* 268–269 (1971).

3 *Bryan, J.H.D. and Umithan, R.R.:* Cytochemical localization of nonspecific esterase and acid phosphatase in spermatozoa of the mouse (*Mus muculus).* Histochemie *33:* 169–180 (1973).

4 *Dott, H.M. and Dingle, J.T.:* Distribution of lysosomal enzymes in the spermatozoa and cytoplasmic droplets of bull and ram. Exp. Cell Res. *52:* 523–540 (1968).

5 *Fritz, H.; Förg-Brey, B.; Fink, E.; Schiessler, H.; Jaumann, E. und Arnhold, M.:*

Charakterisierung einer Trypsin-ähnlichen Proteinase (Akrosin) aus Eberspermien durch ihre Heimbarkeit mit verschiedenen Protein-Proteinase Inhibitoren. I, II, III. Z. Physiol. Chem. *353:* 1007–1014 (1972).

6 *Gould, K.G.:* Application of *in vitro* fertilization. Fed. Proc. *32:* 2069–2074 (1973).

7 *Gould, K.G.; Zaneveld, L.J.D.; Srivastava, P.N., and Williams, W.L.:* Biochemical changes in the zona pellucida of rabbit ova induced by fertilization and sperm enzymes. Proc. Soc. exp. Biol. Med. *136:* 6–10 (1971).

8 *Greenslade, F.C.; McCormack, J.J.; Hirsch, A.F., and Davanzo, J.P.:* Blockage of fertilization in *Rana pipiens* by trypsin inhibitors. Biol. Reprod. *8:* 306–310 (1973).

9 *Ingrisch, H.; Haendle, H. und Werle, E.:* Über die Konzentration des Trypsin-Inhibitors im Sperma von gesunden und andrologisch Kranken und über ihre Beziehung zu anderen Parametern des Spermas. Andrologie *2:* 103–115 (1970).

10 *Kurzrok, R.; Leonard, S.L., and Conrad, H.:* Role of hyaluronidase in human fertility. Amer. J. Med. *1:* 491–506 (1946).

11 *Leonard, S.L. and Kurzrok, R.:* Inhibitors of hyaluronidase in blood sera and their effect on follicle cell dispersal. Endocrinology *39:* 84–90 (1946).

12 *Lewis, B.K. and Ketchel, M.M.:* Effects of female reproductive tract secretions on rabbit sperm. I, II. Proc. Soc. exp. Biol. Med. *141:* 712–724 (1972).

13 *Mancini, R.E.; Alonso, A; Barquet, J.; Alvarez, B., and Nemirovsky, M.:* Histoimmunological localization of hyaluronidase in bull testis. J. Reprod. Fertil. *8:* 325–330 (1964).

14 *Meizel, S.:* Biochemical detection and activation of an inactive form of a trypsin-like enzyme in rabbit testis. J. Reprod. Fertil. *31:* 459–462 (1972).

15 *Meizel, S. and Cotham, J.:* Partial characterization of a new bull sperm arylamidase. J. Reprod. Fertil. *28:* 303–307 (1972).

16 *Metz, C.B.:* Role of specific sperm antigens in fertilization. Fed. Proc. *32:* 2057–2064 (1973).

17 *Metz, C.B.; Seiguer, A.C., and Castro, A.E.:* Inhibition of the cumulus dispersing and hyaluronidase activities of sperm by heterologous and isologous antisperm antibodies. Proc. Soc. exp. Biol. Med. *140:* 776–781 (1972).

18 *Newell, S.O.; Polakoski, K.L., and Williams, W.L.:* Inhibition of fertilization by proteinase inhibitors. 7th Int. Congr. Anim. Reprod. Art. Insem. Münich 1973, vol. 3, pp. 498–502.

19 *Parkes, A.S.; Rogers, H.J., and Spensley, P.C.:* Biological and biochemical aspects of the prevention of fertilization by enzyme inhibitors. Proc. Soc. Study Fertil. *6:* 65–80 (1954).

20 *Pincus, G.; Pirie, N.W., and Chang, M.C.:* The effect of hyaluronidase inhibitors on fertilization in the rabbit. Arch. Biochem. *19:* 388–396 (1948).

21 *Polakoski, K.L.; McRorie, R.A., and Williams, W.L.:* Boar acrosin. I, II. J. biol. Chem. *248:* 8178–8188 (1973).

22 *Polakoski, K.L.; Williams, W.L., and McRorie, R.A.:* Partial purification and characterization of an acidic proteinase in sperm acrosomes. Fed. Proc. *32:* 310 (1973).

23 *Rowlands, I.W.:* Capacity of hyaluronidase to increase the fertilizing power of sperm. Nature, Lond. *154:* 382–383 (1944).

24 *Seiguer, A.C. and Castro, A.E.:* Electron microscopic demonstration of aryl sulphatase activity during acrosome formation in the rat. Biol. Reprod. *7:* 31–42 (1972).

25 *Schumacher, G.F.B.; Swartwout, J.R., and Zuspan, F.P.:* Fertility experiments in mice and rabbits with the trypsin-kallikrein inhibitor from bovine lung; in *Fritz and Tschesche* Proc. Int. Res. Conf. Proteinase Inhib., pp. 247–252 (W. de Gruyter, Berlin 1971).

26 *Schumacher, G.F.B. and Zaneveld, L.J.D.:* Inhibition of rabbit and human sperm acrosomal proteinase by protease inhibitors of human origin. 7th Int. Congr. Anim. Reprod. Art. Insem., Münich 1972, vol. 3, pp. 2105–2110.

27 *Soupart, P. and Clewe, T.H.:* Sperm penetration of rabbit zona pellucida inhibited by treatment of ova with neuraminidase. Fert. Steril. *16:* 677–689 (1965).

28 *Srivastava, P.N. and Gould, K.G.:* Inhibition of fertilizing capacity of rabbit spermatozoa by sialo-proteins. Contraception *7:* 65–73 (1973).

29 *Srivastava, P.N.; Zaneveld, L.J.D., and Williams, W.L.:* Mammalian sperm acrosomal neuraminidases. Biochem. biophys. Res. Commun. *39:* 575–582 (1970).

30 *Stambaugh, R. and Buckley, J.:* Identification and subcellular localization of the enzymes effecting penetration of the zona pellucida by rabbit spermatozoa. J. Reprod. Fertil. *19:* 423–432 (1969).

31 *Stambaugh, R.; Seitz, M., and Mastroianni, L.:* Acrosomal proteinase inhibitors in rhesus monkey *(Macaca mulatta)* oviduct fluid. Biol. Reprod. *9:* 86 (1973).

32 *Suominen, J.; Kaufman, M.H., and Setchell, B.P.:* Prevention of fertilization *in vitro* by an acrosin inhibitor from rete testis fluid of the ram. J. Reprod. Fertil. *34:* 385–388 (1973).

33 *Suominen, J. and Setchell, B.P.:* Enzymes and trypsin inhibitor in the rete testis fluid of rams and boars. J. Reprod. Fertil. *30:* 235–245 (1972).

34 *Tauber, P.F.; Zaneveld, L.J.D.; Propping, D., and Schumacher, G.F.B.:* Composition of human split ejaculates. II. Enzymes, activator and inhibitors (in preparation).

35 *Teichman, R.J. and Bernstein, M.H.:* Fine structure localizations of acid phosphatase in rabbit and bull sperm heads. J. Reprod. Fertil. *27:* 243–248 (1971).

36 *Williams, W.L.; Robertson, R.T., and Dukelow, W.R.:* Decapacitation factor and capacitation. Adv. Biosci. *4:* 61–72 (1970).

37 *Yanagimachi, R. and Teichman, R.J.:* Cytochemical demonstration of acrosomal proteinase in mammalian and avian spermatozoa by a silver proteinate method. Biol. Reprod. *6:* 87–97 (1972).

38 *Yang, S.-L.; Zaneveld, L.J.D., and Schumacher, G.F.B.:* Effect of serum proteinase inhibitors on the fertilizing capacity of rabbit spermatozoa (in preparation).

39 *Zaneveld, L.J.D.; Dragoje, B.M., and Schumacher, G.F.B.:* Acrosomal proteinase and proteinase inhibitor of human spermatozoa. Science *177:* 702–703 (1972).

40 *Zaneveld, L.J.D.; Polakoski, K.L., and Schumacher, G.F.B.:* Properties of acrosomal hyaluronidase from bull spermatozoa. Evidence for its similarity to testicular hyaluronidase. J. biol. Chem. *248:* 564–570 (1973).

41 *Zaneveld, L.J.D.; Polakoski, K.L.; Robertson, R.T., and Williams, W.L.:* Trypsin inhibitors and fertilization; in *Fritz and Tschesche* Proc. Int. Res. Conf. Proteinase Inhib., pp. 236–244 (W. de Gruyter, Berlin 1971).

42 *Zaneveld, L.J.D.; Polakoski, K.L., and Williams, W.L.:* Properties of a proteolytic enzyme from rabbit sperm acrosomes. Biol. Reprod. *6:* 30–39 (1972).

43 *Zaneveld, L.J.D.; Polakoski, K.L., and Williams, W.L.:* A proteinase and proteinase inhibitor of mammalian sperm acrosomes. Biol. Reprod. *9:* 219–225 (1973).

44 *Zaneveld, L.J.D.; Schumacher, G.F.B., and Travis, J.:* Human sperm acrosomal proteinase. Antibody inhibition and immunologic dissimilarity to human pancreatic trypsin. Fertil. Steril. *24:* 479–484 (1973).

45 *Zaneveld, L.J.D. and Williams, W.L.:* A sperm enzyme that disperses the corona radiata and its inhibition by decapacitation factor. Biol. Reprod. *2:* 363–368 (1970).

Author's address: Prof. *L.J.D. Zaneveld,* Population Research Center, Medical Sciences and Engineering Division, IIT Research Institute, 10 West 35th Street, *Chicago, IL 60616* (USA)

Practical Aspects

The Biology of Spermatozoa
INSERM Int. Symp., Nouzilly 1973, pp. 202–213 (Karger, Basel 1975)

Contraception and Sperm Transport in Domestic Animals

T.J. Robinson
Department of Animal Husbandry, University of Sydney, Sydney

Introduction

Failure of fertilization is the prime cause of lowered fertility following two major manipulative processes, namely the long-term storage of semen and the control of ovulation. This failure can be avoided by the use of excessive numbers of spermatozoa or by their surgical introduction near the utero-tubal junction. Failure is due to an impairment of normal transport and survival of spermatozoa.

Potential Barriers to Sperm Transport

The vaginal environment is probably relatively unimportant except in cases of acute infection or inflammation, as the site of deposition of the sperm is coated with cervical secretions.

The cervix is important both as a potential barrier and as a reservoir in those species such as man, the cow, sheep and goat in which vaginal insemination is the rule. In other species such as the pig and horse, in which the ejaculate enters the uterus, the role of the cervix is not so clear.

The uterus, because of its size, must be a potential barrier and conditions such as infection, incomplete resolution after parturition, and the like, must affect transport.

The utero-tubal junction and the Fallopian tubes provide another potential hazard. The environment in the tubes at about the time of fertilization differs from that of the uterus (38).

There are no accurate data on the numbers of spermatozoa needed in the tubes for a high chance of fertilization, but several thousand appear necessary.

Numbers build up gradually over 24 h in the sheep, and this involves survival in four distinct anatomical and potentially different physiological regions, and the establishment of one or more protected reservoirs.

Location of Sperm Reservoirs and their Establishment

Species with Vaginal Insemination

The cervix acts as the main reservoir, in the sheep, cow and goat (32–34, 40), and spermatozoal motility is essential (36). The reservoir is established within 1 h in the ewe and may be maintained at a high level for 24–36 h (42). Penetration through the cervix is rapid so that substantial numbers are present in the cranial third 2 h after insemination (14). Passage from the cervix into the uterus follows rapidly upon the establishment of the population in the cranial third and the cervical mucus may be involved (30).

The utero-tubal junction and the uterine glands provide a second reservoir in the cow (52) and ewe (42, 53). *Thibault* (52) believes that the cervical reservoir is important only during the first 24 h and that thereafter the utero-tubal junction is the reservoir responsible for replenishment of the tubes.

Species with Uterine Insemination

The utero-tubal junction appears to be the site of the main reservoir in the sow (10, 39, 45, 46). Spermatozoal motility is not necessary for transport although it may be involved in the actual penetration of the reservoirs (6).

Endocrinological and Other Factors Involved in Establishment of Sperm Reservoirs

Maximum chances of conception are provided when disease free, well-nourished animals are mated with highly fertile males in a stress-free environment at the height of the breeding season when endocrine status is at a maximum for reproductive efficiency. Such conditions favour a high percentage of animals fertilized largely as a function of efficiency of sperm transport.

Disease

The effects on sperm transport of venereal disease of the male and of inflammatory diseases of the female tract do not come into the scope of this paper.

Nutrition

The effects of malnutrition on the production of spermatozoa in the male are well documented. By contrast, the effects on sperm transport in the female are not. It is a well-recognized field observation that poor conception rates to artificial insemination of sheep can be expected under drought conditions. This probably involves sperm transport and fertilization.

Stress

Stress reduces the numbers of spermatozoa which reach the upper part of the uterine horns and the Fallopian tubes (32, 35, 54) and impairs passage through the cervix (44, 53).

Unfamiliar surroundings, the presence of dogs, and even the act of artificial insemination (20) may provide a stressful situation. In view of the inhibition of oxytocin-induced contractions of the uterus by adrenaline (27) one can deduce, despite contrary evidence (37), that muscular contractions are involved in penetration through the cervix and that stress reduces this activity.

Endocrine Status

Indirect evidence for the ewe indicates that the annual rhythm of breeding activity is associated with a corresponding rhythm of pituitary gonadotrophin output and ovarian activity (16–18, 31, 56).

Direct confirmation of this is now available as a consequence of the use of competitive protein binding and radioimmunoassay techniques and it is significant that 'maximum reproductive efficiency', of which sperm transport is an essential part, is associated with high efficiency of output of the trophic and target hormones and of the feedback and other associated mechanisms. The local environment of the female reproductive tract is a function of activity of the ovarian hormones, so the level of these hormones must be important in sperm transport.

For forty years it has been known that conception does not follow forced matings of ewes after injection of PMSG in anoestrus, despite apparently normal ovulation (11). Progesterone insufficiency in the inter-oestrous period has been implicated in the subfertility associated with impaired sperm transport following progestagen treatment (4, 42, 43, 47). Further, the amount of oestrogen used to induce oestrus in ovariectomized ewes affects the number of spermatozoa in the cervix following insemination (table I) (1–3, 13). By contrast with the situation in stressed ewes, there is no evidence of a specific block within the cervix.

It is not known whether the effect of inadequate progestagen before oestrus is (a) direct, due to inadequate preparation of the tract (41), or (b) indirect, due to a change in the ovarian hormone pituitary feedback mechanism. The level of oestrogen produced is affected by inadequate progestagen (51) as is the nature and quantity of the cervical mucus (13, 50).

Table I. Mean numbers of spermatozoa in the cervix of progesterone-oestrogen-treated spayed ewes (13)

Interval from insemination to slaughter, h	Dose of oestradiol benzoate used to induce oestrus, µg	Region of cervix		
		caudal	mid (× 10^3)	cranial
1	30	2,680	50	3
	90	4,730	552	6
12	30	135	71	17
	90	373	1,090	181
24	30	85	135	31
	90	176	166	63

Table II. Spermatozoa recovered from the cervix and uterus 2 h after insemination and the Fallopian tubes at 24 h and the relationship with fertility of ewes treated with oestrogen during dioestrus (14, 48)

	Mean numbers of spermatozoa					Fertility
	cervix			uterus (× 10^3)	tubes	percent ewes pregnant
	caudal	mid (× 10^3)	cranial			
A. Cronolone treated						
No oestrogen	4,780	800	42	4.4	127	58
1 µg OD/day	1,570	650	66	0.3	25	61
5 µg OD/day	1,610	240	8	0.1	< 10	30
B. Normal cycle						
No oestrogen	820	190	37	0.2	100	49
1 µg OD/day	980	310	41	0.8	60	49
5 µg OD/day	1,260	500	42	0.1	100	46
25 µg OD/day	490	16	0.6	0.02	< 10	6

Exogenous Oestrogen

Failure of sperm transport is involved in the subfertility observed in ewes grazed on oestrogenic pastures (28). A small daily dose of oestrogen administered during the dioestrous period preceding either a normal or controlled oestrus also impairs sperm transport (table II). As in the case of stress, the cervix seems to act as a specific barrier.

Mechanisms of Transport in Different Regions of the Reproductive Tract

In the sow and the mare, the large volume of ejaculate enters the uterus and is spread through the tract so that uterine and utero-tubal junction reservoirs are established rapidly.

In the cow, ewe and goat-doe, there are at least four stages of sperm transport, namely (1) entry into the cervix, (2) passage through the cervix, (3) passage through the uterus, and (4) entry into and through the tubes. The first two stages appear primarily dependent on spermatozoal motility (29, 36) although the role of muscular contraction cannot be ignored. The third depends upon uterine contractions (6, 49) and the fourth upon a combination of spermatozoal motility and tubal contractions (15).

Entry into the Cervix

Entry into the caudal third of the cervix is extremely rapid; large numbers may be recovered within 1 or 2 h and, when subjected to log transformation, constitute a normal distribution. Treatments known to suppress fertility do not block entry (14).

Passage through the Cervix

Passage through the cervix is rapid and is oestrogen dependent (2, 3, 5) with dose and method of administration important. Further, there is an interaction with the preceding progesterone regime. Passage may be severely impaired by oestrogen influence during the progestational phase preceding oestrus, and there is evidence of a specific barrier within the cervix.

The pattern for subsequent fertility appears to be established within 2 h of insemination. The numbers of spermatozoa present in the cranial third of the cervix at 2 h after insemination predict the numbers in the tubes at 24 h and consequent fertility (table II).

Table III. Positive correlations (r) between numbers of spermatozoa in the tract of ewes 2 and 24 h after insemination (14)

Oestrus	Interval from insemination h	Region of tract			
		cervix		cranial/ uterus	uterus/ tubes
		caudal/mid	mid/cranial		
Normal	2	0.70	0.69	0.50	nil
	24	0.78	0.84	0.67	0.78
Controlled	2	0.65	0.84	0.22	nil
	24	0.69	0.82	0.50	0.47

Passage through the Uterus

Substantial numbers of spermatozoa may be present in the uterus at 2 h (14, 52), and the numbers are positively correlated with those found in the cranial third of the cervix (table III). Few, if any, are yet to be found in the tubes.

There is no direct evidence of a specific effect of the uterus deleteriously affecting passage of spermatozoa but there is some indirect evidence (41, 42).

Passage into the Tubes

Muscular activity and motility are involved (52). Dead spermatozoa appear transported through the utero-tubal junction of sows as rapidly as live and there is evidence of a barrier (6). In the normal ewe the numbers recoverable from the Fallopian tubes at 24 h are well correlated with the numbers in the uterus (table III).

Modification of Transport and Storage

Modification of Endocrinological Environment

The Cervix

In those species with vaginal insemination, adverse situations such as stress, anaesthesia (29) and oestrogen influence during dioestrus, may result in a total cervical block, with two factors involved, namely modification of the cervical mucus and alteration of the pattern of muscular contractions.

Table IV. Numbers of spermatozoa recovered from the tracts of cyclic and progestagen-treated ewes 24 h after insemination (25)

Treatment and condition of spermatozoa	Region of tract			
	vagina	cervix	uterus	tubes
Normal cycle				
Intact sperm	263,000	59,800	2,050	249
Tailless sperm	171,000	11,200	905	136
Tailless, %	39	16	31	35
Progestagen sponge				
Intact sperm	3,390	4,360	741	62
Tailless sperm	8,910	3,980	525	85
Tailless, %	70	46	41	58

Modification of the mucus is involved in most, if not all, cases of endocrinological change associated with impaired sperm transport but results are conflicting and variable (13, 44, 50). These studies have dealt with marginal changes in endocrine status which could fall within normal physiological limits but which still affect sperm transport. The use of the progestagen-oestrogen interaction in the entire and in the spayed ewe offers a most attractive model for critical studies in sperm transport and contraception. By appropriate manipulation of the endocrine environment one can regulate conception rates from 0 to 76 % (48).

The Uterus

Spermatozoa disappear more rapidly from the uteri of ewes under the influence of oestrogen than of progesterone, and this is associated with excessive sperm 'breakage' (12, 26). Similarly, intensification of sperm 'breakage' and disappearance is found in ewes in which the time of ovulation is controlled with progestagen-impregnated intravaginal sponges (23, 25). There is a reduction in the total numbers of spermatozoa recoverable from the vagina, cervix, uterus and tubes of treated ewes, as compared with normal cyclic animals, and a marked increase in the percentage of tailless or 'broken' spermatozoa (table IV). These data do not implicate the uterus as a *specific* barrier – all regions of the tract are involved.

Modification of Physical Environment

Hawk and his associates have studied the fate of spermatozoa in the reproductive tract following manipulation of the physical as well as the endocrinological environment. In 1964 they published the first of a series of papers on the contraceptive effect of intrauterine devices (IUD) in sheep and cattle (24). The failure of sperm transport in the contra-lateral as well as the ipsi-lateral horn of the uterus was soon recognized as the probable cause of failure of fertilization (21). Later, this was associated with 'breakage' of spermatozoa into heads and tails (22) and with an increase in amplitude and a reversal in direction of normal uterine contractions (8, 9). Attempts to modify the antifertility effect of the IUD by modification of the uterine pattern of contractility failed (55). Accelerated phagocytosis is implicated as the cause of disappearance of spermatozoa and the uterus of an IUD-treated ewe provides a greater leucocytic response to semen than does that of a normal ewe (7, 12).

Experimental Techniques

Thibault concludes that the only infallible technique for the study of sperm migration is fixation of the tract *in situ* or immediately after slaughter, followed by serial section and subsequent counting of spermatozoa. The amount of work involved and the small number of animals one can use pose serious limiting factors. Other histological techniques such as those of *Hafez* (19) are elegant for detailed studies on individuals but are not suitable for quantitative experimentation.

On the other hand, techniques involving the flushing out of regions of the tract and subsequent counting of spermatozoa, while permitting the handling of more animals, and hence of experimental treatment combinations, are open to the following serious criticisms.

(a) The numbers of spermatozoa in the distal regions of the tract are so small relative to those in the caudal regions that the chances of contamination are great. With care, this problem can be avoided, so that recent data on the rate of sperm transport derived from flushing experiments are in accord with those obtained following fixation.

(b) The numbers recovered from each region of the tract are highly variable and, in the case of the uterus and tubes, there is a high proportion of zero values. This is disturbing from two points of view. First, no sperm may be found in washings of tubes from which eggs are flushed with adherent spermatozoa; clearly these were not the only spermatozoa present in the tubes. Second, the data pose serious statistical problems. They are highly skew with occasional very high values, which necessitates a log transformation, and one must add a constant to deal with the zero values.

(c) *Mattner* (34) found that, despite the flushing out of about 90 % of the cervical mucus of goats and cows 18–24 h after mating, only about 50 % of the spermatozoa were recovered. Hence, data are only of value if related to other data collected in the same experiment.

In our laboratory, we do not relate numbers of spermatozoa counted in one experiment with those counted in another. However, within experiments, we do obtain meaningful comparisons, and the pattern in one experiment can be repeated in another. Variability is a problem, particularly when it is compounded with a high proportion of zero values. However, there are cases where the variability – particularly if associated with significant skewness or kurtosis – may have biological significance and the definition of its nature may assist in interpretation of data (42). Concerning zero values, the demonstration of a relationship between numbers of spermatozoa in the cranial third of the cervix at 2 h with the numbers in the tubes at 24 h after insemination is a most useful observation. Of 64 cervices examined 2 h after insemination, none yielded a zero value for the caudal third and only 2 and 7 for the mid and cranial thirds. The distributions following $\log_{10}$ transformation were, for the most part, normal and capable of formal statistical analysis. Second, departure from normality, including the appearance of zero values in the cranial third of the cervix, was found only in those ewes in which treatment had impaired sperm transport, and so has biological significance.

For comparative work on the effect of treatments on sperm transport, particularly through the cervix of large animals, the sperm flushing technique is extremely valuable and the results are capable of analysis. For detailed work on the nature of sperm transport, loss, and the like, the more exacting techniques of fixation and section are required. The different techniques, properly used, must be regarded as complementary, and their respective limits and merits recognized.

References

1 *Allison, A.J.:* The effect of exogenous hormones on reproductive phenomena in the ewe; Thesis, Sydney (1970).

2 *Allison, A.J.:* Transport of spermatozoa in spayed ewes: comparison of two methods of steroid administration. J. Reprod. Fertil. *31:* 309–312 (1972a).

3 *Allison, A.J.:* Comparison of the transport of spermatozoa in spayed and entire ewes. J. Reprod. Fertil. *31:* 415–423 (1972b).

4 *Allison, A.J. and Robinson, T.J.:* The effect of dose level of intravaginal progestagen on sperm transport, fertilization and lambing in the cyclic Merino ewe. J. Reprod. Fertil. *22:* 515–531 (1970).

5 *Allison, A.J. and Robinson, T.J.:* Recovery of spermatozoa from the reproductive tract of the spayed ewe treated with progesterone and oestrogen. J. Reprod. Fertil. *31:* 215–224 (1972).

6 *Baker, R.D. and Degan, A.A.:* Transport of live and dead spermatozoa within the reproductive tract of gilts. J. Reprod. Fertil. *28:* 369–377 (1972).

7 *Brinsfield, T.H.; Clark, M.V., and Hawk, H.W.:* Phagocytosis as the basis of accelerated sperm disappearance in ligated IUD-bearing uterine horns of the ewe. Biol. Reprod. *5:* 78–81 (1971).

8 *Brinsfield, T.H. and Hawk, H.W.:* Stimulation by intra-uterine devices of motility of sheep myometrium. J. Reprod. Fertil. *16:* 129–131 (1968).

9 *Brinsfield, T.H. and Hawk, H.W.:* Modification of the direction of uterine contractions by intra-uterine devices in the ewe. J. Reprod. Fertil. *18:* 535–537 (1969).

10 *Mesnil du Buisson, F. du et Dauzier, L.:* Distribution et résorption du sperme dans le tractus génital de la truie: survie des spermatozoïdes. Ann. Endocrin., Paris *16:* 413–422 (1955).

11 *Cole, H.H.; Hart, G.H., and Miller, R.F.:* Studies on the hormonal control of estrus phenomena in the anestrous ewe. Endocrinology, Springfield *36:* 370–380 (1945).

12 *Conley, H.H. and Hawk, H.W.:* Intensification by intrauterine devices of sperm loss from the sheep uterus. Biol. Reprod. *2:* 401–407 (1970).

13 *Croker, K.P.:* Studies on the causes of the relative infertility observed following control of oestrus in the sheep; Thesis, Sydney (1973).

14 *Croker, K.P.; Robinson, T.J., and Shelton, J.N.:* The passage of spermatozoa through the reproductive tract of the ewe (in press).

15 *Dauzier, L.:* Physiologie du déplacement des spermatozoïdes dans les voies génitales femelles chez la brebis et la vache; Thesis, Paris (1958).

16 *Fletcher, I.C.:* Interrelationships between hormones, behavior and fertility in sheep; Thesis, Sydney (1968).

17 *Fletcher, I.C. and Lindsay, D.R.:* Effect of oestrogen on oestrous behaviour and its variation with season in the ewe. J. Endocrin. *50:* 685–696 (1971).

18 *Grant, R.:* Studies on the physiology of reproduction in the ewe. Trans. roy. Soc. Edinb. *58:* 1–47 (1934).

19 *Hafez, E.S.E.:* Histology and microstructure of the cervical epithelial secretory system. WHO Coll. on Cervical Mucus in Human Reproduction, Geneva 1972.

20 *Hancock, J.L. and McGovern, P.T.:* The transport of sheep and goat spermatozoa in the ewe. J. Reprod. Fertil. *15:* 289–294 (1968).

21 *Hawk, H.W.:* Inhibition of ovum fertilization in the ewe by intra-uterine plastic spirals. J. Reprod. Fertil. *10:* 267–269 (1965).

22 *Hawk, H.W.:* Investigations into the anti-fertility effect of intra-uterine devices in the ewe. J. Reprod. Fertil. *14:* 49–59 (1967).

23 *Hawk, H.W.:* Progestagen induced sperm breakage in the sheep vagina. J. anim. Sci. *34:* 795–798 (1972).

24 *Hawk, H.W.; Conley, H.H.; Brinsfield, T.H., and Richter, H.F.:* Contraceptive effect of plastic devices in cattle uteri. Proc. 2nd Int. Conf. on Intra-Uterine Contraception (Excerpta Medica, New York 1964).

25 *Hawk, H.W. and Conley, H.H.:* Loss of spermatozoa from the reproductive tract of the ewe and intensification of sperm 'breakage' by progestagen. J. Reprod. Fertil. *27:* 339–347 (1971).

26 *Hawk, H.W. and Cooper, B.S.:* Influence of ovarian hormones on sperm 'breakage' and disappearance in sheep uteri. J. Reprod. Fertil. *27:* 349–354 (1971).

27 *Hays, R.L. and Demark, N.L. van:* Effect of oxytocin and epinephrine on uterine motility in the bovine. Amer. J. Physiol. *183:* 510–512 (1953).

28 *Lightfoot, R.J.; Croker, K.P., and Neil, H.G.:* Failure of sperm transport in relation to ewe infertility following prolonged grazing on oestrogenic pastures. Austr. J. agric. Res. *18:* 755–765 (1967).

29 *Lightfoot, R.J. and Restall, B.J.:* Effects of site of insemination, sperm motility and

genital tract contractions on transport of spermatozoa in the ewe. J. Reprod. Fertil. *26:* 1–13 (1971).

30 *Lineweaver, J.A.; Hafez, E.S.E.; Ehlers, M.H.; Dickson, W.M., and King, J.R.:* Sperm transport in cattle and gonadotrophin-treated calves. Cornell Vet. *60:* 372–382 (1970).

31 *McKenzie, F.F. and Terrill, C.E.:* Estrus, ovulation and related phenomena in the ewe. Res. Bull. Mo. agric. exp. Sta. *264* (1937).

32 *Mattner, P.E.:* Spermatozoa in the genital tract of the ewe. II. Distribution after coitus. Austr. J. Biol. Sci. *16:* 688–694 (1963).

33 *Mattner, P.E.:* Formation and retention of the spermatozoan reservoir in the cervix of the ruminant. Nature, Lond. *212:* 1479–1480 (1966).

34 *Mattner, P.E.:* The distribution of spermatozoa and leucocytes in the female genital tract in goats and sheep. J. Reprod. Fertil. *17:* 253–261 (1968).

35 *Mattner, P.E. and Braden, A.W.H.:* Spermatozoa in the genital tract of the ewe. I. Rapidity of transport. Austr. J. Biol. Sci. *16:* 473–481 (1963).

36 *Mattner, P.E. and Braden, A.W.H.:* Comparison of the distribution of motile and immotile spermatozoa in the ovine cervix. Austr. J. Biol. Sci. *22:* 1069–1070 (1969).

37 *Moghissi, K.S.:* Sperm migration through the human cervix. WHO Coll. on Cervical Mucus in Human Reproduction, Geneva 1972, pp. 128–152.

38 *Moore, N.W. and Shelton, J.N.:* Egg transfer in sheep. Effect of degree of synchronization between donor and recipient, age of egg, and site of transfer on the survival of transferred eggs. J. Reprod. Fertil. *7:* 145–152 (1964).

39 *Pitkjanen, I.J.:* Recent advances in fertilization and increased fecundity in pigs (Sel' kozgiz, Moscow 1961).

40 *Quinlan, J.; Maré, G.S., and Roux, L.:* The vitality of the spermatozoon in the genital tract of the Merino ewe with special reference to its practical application in breeding. Rep. Vet. Serv. Anim. Ind. Sth Afr. *18:* 831–870 (1932).

41 *Quinlivan, T.D.:* Studies in ovine reproduction; Thesis, Sydney (1967).

42 *Quinlivan, T.D. and Robinson, T.J.:* The number of spermatozoa in the Fallopian tubes of ewes at intervals after artificial insemination following withdrawal of SC-9880-impregnated intravaginal sponges; in The control of the ovarian cycle in sheep, pp. 177–194 (Sydney University Press, Sydney 1967).

43 *Quinlivan, T.D. and Robinson, T.J.:* Numbers of spermatozoa in the genital tract after artificial insemination of progestagen-treated ewes. J. Reprod. Fertil. *19:* 73–86 (1969).

44 *Raynaud, F.:* Physiologie du transport des spermatozoïdes dans le col utérin de la brebis; Thèse, Paris (1972).

45 *Rigby, J.P.:* The fate of spermatozoa in the genital tract of the sow. Proc. 5th Int. Congr. of Animal Production, Trento 1964, vol. 4, pp. 421–425.

46 *Rigby, J.P.:* The persistence of spermatozoa at the utero-tubal junction of the sow. J. Reprod. Fertil. *11:* 153–155 (1966).

47 *Robinson, T.J.:* The control of the ovarian cycle in the sheep (Sydney University Press, Sydney 1967).

48 *Robinson, T.J.:* Factors involved in the failure of sperm transport and survival in the female reproductive tract. J. Reprod. Fertil. *18:* 103–109 (1973).

49 *Rowson, L.E.A.:* The movement of radio opaque material in the bovine uterine tract. Brit. vet. J. *111:* 334–339 (1955).

50 *Smith, J.F. and Allison, A.J.:* The effect of exogenous progestagen on the production of cervical mucus in the ewe. J. Reprod. Fertil. *24:* 279–282 (1971).

51 *Smith, J.F. and Robinson, T.J.:* The effect of exogenous progestagen on the levels of

free oestrogen in the ovarian vein plasma of the ewe. J. Endocrin. *48:* 485–496 (1970).

52 *Thibault, C.:* Sperm transport and storage in vertebrates. J. Reprod. Fertil. *18:* 39–53 (1973).

53 *Thibault, C. and Wintenberger-Torres, S.:* Oxytocin and sperm transport in the ewe. Int. J. Fertil. *12:* 410–415 (1967).

54 *Thibault, C.; Wintenberger-Torres, S. et Raynaud, F.:* Influence du stress et des progestogènes de synthèse sur le transport des spermatozoïdes chez la brebis; cited by *Raynaud* (44).

55 *Warren, J.E. and Hawk, H.W.:* Attempts to overcome the antifertility effect of intrauterine plastic spirals in the ewe. J. anim. Sci. *27:* 131–133 (1968).

56 *Watson, R.H.:* Seasonal variation in reproductive activity in ewes. Austr. vet. J. *28:* 1–5 (1952).

Author's address: Prof. *T.J. Robinson,* Department of Animal Husbandry, The University of Sydney, *Sydney, NSW 2006* (Australia)

The Biology of Spermatozoa
INSERM Int. Symp., Nouzilly 1973, pp. 214–223 (Karger, Basel 1975)

Sperm Transport in the Ewe: The Effect of Ovarian Steroids

A.J. Allison
Invermay Agricultural Research Centre, Mosgiel

I. Introduction

Recent information concerning the cause of lowered fertilization rates in ewes following progestagen treatment (1) has shown an earlier peak of oestrogen production relative to the onset of oestrus in progestagen-treated ewes in comparison with untreated animals (2). These differences can almost certainly be implicated in the impaired transport and survival of spermatozoa in the treated ewes (3). The obvious need for a greater understanding of the relationships between sperm transport and endocrine status in the ewe prompted the following experimental work. Detailed experimental methods, data and analysis of the following experiments have been presented elsewhere (4, 5).

II. Methods and Results

Experiment 1

Differing doses of progesterone and oestradiol benzoate (ODB) were administered to spayed ewes in an effort to partition the effect of these hormones on sperm transport in the ewe and to define any interaction between the two.

The design was a 4 × 4 factorial with five animals per group.

Dose of progesterone, mg/day	0	v	5	v	10	v	20
Dose of ODB, μg	0	v	12.5	v	25	v	50

The progesterone was administered daily for 12 days, followed 48 h later by a single injection of ODB. Ewes were inseminated with 0.15 ml of undiluted semen, containing not less than 300 × 10^6 spermatozoa, 24 h after injection of ODB. Animals were slaughtered 24 h after insemination and spermatozoa were recovered from all sections of the reproductive tract (4).

Fourteen ewes only yielded spermatozoa from the uterus and 14 from the Fallopian tubes, three ewes had spermatozoa in both sections of the tract. There was no relationship between hormone treatment and recovery.

Estimated numbers of spermatozoa recovered from the vagina and cervix were highly variable and were subjected to $\log_{10}$ transformation prior to analysis. There was no relationship between the numbers of spermatozoa recovered from the vagina and from the cervix. Data for numbers recovered from the cervix are summarized in tables I and II. Oestrogen was necessary for the maintenance of a population of spermatozoa in the cervix until 24 h after insemination and within the oestrogen-treated ewes there was a trend towards larger numbers with an increasing dose of ODB ($p \simeq 0.1$).

The proportion of ewes which yielded large numbers of spermatozoa increased with an increasing dose of oestrogen ($p < 0.01$) (4). Progesterone had no average effect on the mean numbers of spermatozoa but was involved in an interaction with oestrogen.

Table I. Experiment 1: Estimated numbers of spermatozoa recovered from the cervix (× 10^3). Main effects

Progesterone mg/day	Numbers of spermatozoa	Oestrogen μg	Numbers of spermatozoa
0	14.7	0	2.0
5	27.1	12.5	20.0
10	47.8	25.0	41.2
20	46.8	50.0	73.2

Progesterone, not significant.
Oestrogen, 0 v Σ 12.5, 25.0, 50.0 ($p < 0.01$).

Table II. Experiment 1: Estimated numbers of spermatozoa recovered from the cervix (× 10^3). Significant interaction, linear progesterone × linear oestrogen ($p < 0.01$)

Oestrogen, μg	Progesterone, mg/day		
	5	10	20
12.5	59.3	7.4	5.4
25.0	34.0	72.0	46.0
50.0	14.7	108.8	132.6

Table III. Experiment 2: Estimated numbers of spermatozoa recovered from various divisions of the reproductive tract

Main effect	Division of reproductive tract			
	vagina ($\times 10^6$)	cervix ($\times 10^6$)	uterus ($\times 10^3$)	Fallopian tubes ($\times 10^3$)
Type of ewe				
Spayed	35.1	7.5	7	1
Entire	88.5**	27.5***	84*	13***
Time of recovery				
2 h	123.3	30.8	23	2
24 h	0.3***	4.3***	68	12***
Type of mating				
Artificial insemination	47.8	18.6	55	0.6
Natural mating	75.9	16.5*	36	8.3***
Position within cervix				
Caudal		10.5		
Mid		5.8		
Cranial		1.4***		

* $p < 0.05$; ** $p < 0.01$; *** $p < 0.001$.

Table IV. Experiment 2: Estimated numbers of spermatozoa recovered from the cervix at different times after mating. Significant interaction. Position within cervix × time of recovery ($p < 0.001$)

Time of recovery	Position within cervix		
	caudal	mid	cranial
2 h	20.2	9.6	1.2
24 h	0.7	1.9	1.6

Table V. Experiment 2 Estimated numbers of spermatozoa recovered from the cervix of spayed and entire ewes. Significant interaction. Type of ewe × position within cervix ($p < 0.01$)

Type of ewe	Position within cervix		
	caudal	mid	cranial
Spayed	6.1	1.9	0.2
Entire	15.1	9.7	2.7

There was a highly significant linear progesterone by linear oestrogen interaction ($p < 0.01$, table II). Pretreatment with higher doses of progesterone appeared necessary for maximum expression of the effect of oestrogen.

Experiment 2

The numbers of spermatozoa recovered from the spayed ewes in experiment 1 were much lower than those reported from entire ewes (3, 6). Contrary to these findings, *Mattner and Braden* (6) recorded similar numbers of spermatozoa in spayed and entire ewes and also found no effect of stilboestrol injections on numbers recovered from spayed ewes following natural or forced mating. Their data however, relate to recovery $3^1/_2$ to 4 h after mating whereas in experiment 1 recovery was 24 h after artificial insemination. Consequently, the transport of spermatozoa in spayed and entire ewes was compared at two times of recovery following artificial insemination or natural mating. Five ewes per group were used and spermatozoa were recovered 2 or 24 h after mating. Following slaughter cervices were snap frozen in liquid nitrogen vapour (5) allowing greater subsequent recovery than previously reported techniques (6, 20). The results are presented in tables III, IV and V.

Spayed ewes had significantly fewer spermatozoa than entire ewes in all divisions of the reproductive tract. Fewer spermatozoa were in the vagina and cervix at 24 h while the reverse was the case for the Fallopian tubes. Populations of spermatozoa recovered from the uterus were not different at the two times of recovery. Natural mating to each of two rams within a 5-min period did not result in a greater population of spermatozoa in the vagina 2 h after mating in comparison with artificial insemination. Obviously, a large portion of the inseminate is lost – probably voided to the exterior within 2 h of mating.

Within the cervix there was a highly significant main effect showing a marked gradation in numbers of spermatozoa from the caudal to the cranial end but the distribution was different between the two times of recovery. Numbers recovered from the cranial portion of the cervix were similar at both times of recovery whereas mid and caudal populations dropped sharply between 2 and 24 h ($p < 0.001$, table IV).

The distribution of spermatozoa within the cervix was also greatly different between spayed and entire ewes ($p < 0.01$, table V). In the former a much lower proportion of the initial cervical population reached the cranial and mid thirds and was maintained there until recovery 24 h after insemination. In fact the mean number recovered from the cervices of spayed ewes 24 h after mating was only 5 % of the number recovered at 2 h, in comparison with a figure of 16 % for entire ewes.

Histological examination of the uterine endometrium indicated an early stage of atrophy in the spayed ewes (5) even though they had been injected with progesterone from the day of ovariectomy and then oestrogen. Thus, injection schedules used did not maintain the uterus in a normal physiological state.

III. Discussion

The preceding data provide information on endocrine interaction in the transport of spermatozoa and also basic information concerning the pattern of transport within the reproductive tract. Since recovery of spermatozoa in experiment 1 was carried out only 24 h after insemination it may be argued that the observed pattern of response to the endocrine treatments is not necessarily a reflection of the pattern of initial penetration and establishment of a cervical population. The data therefore, simply show that oestrogen is directly involved in the initial establishment of a population within the cervix and its maintenance for 24 h without differentiating between establishment and maintenance.

Changes in uterine motility and the characteristics of cervical mucus have been implicated in the transport of spermatozoa and both are under the influence of ovarian steroids (7, 8) maximal responses coinciding with production of endogenous oestrogen. Although the uterus of the spayed ewe shows moderate contractile activity without hormonal influences the activity is depressed or enhanced by injections of progesterone or oestrogen, respectively (9). Detailed studies on the structure of human cervical mucus (10) have shown the parallel arrangement of macromolecular micelles characteristic of the oestrogenic phase of the cycle, to breakdown in the luteal phase when the micelles split up forming a dense network which does not allow the penetration of spermatozoa. Although uterine motility is minimal in the absence of oestrogen it is most likely that the low numbers of spermatozoa recovered from the cervix in ewes which did not receive oestrogen were primarily due to the inability of spermatozoa to penetrate thick viscous mucus.

The significant linear oestrogen by linear progesterone interaction implies that a balance of steroids is involved in the establishment and maintenance of a population of spermatozoa within the cervix. Imbalance may affect motility of the reproductive tract, in particular the uterus. Evidence for the involvement of myometrial activity in the transport of spermatozoa is conflicting. In normal ewes halothane anaesthesia, a potent inhibitor of spontaneous contractions, has no effect on the population of spermatozoa recovered from the cervix 30 min after insemination (11). However, reversal of the directional propagation of uterine contractions is almost certainly involved in the impaired transport of spermatozoa caused by intra-uterine devices in ewes (12, 13).

Synchronization of oestrus by administration of progestagens orally, subcutaneously or intravaginally has also been shown to cause a similar change in

Table VI. Directional classification of uterine contractions in spayed and entire ewes. From *Croker and Shelton* (15)

	Dose of ODB, μg	Contractions number/min	Percentage of contractions			
			towards tubes	towards body	stationary	indefinite
Spayed ewes	10	2.7	5.7	73.4	14.8	6.1
	30	2.8	38.7	24.8	14.2	22.3
	90	3.5	33.1	11.4	9.4	46.0
Entire ewes (early oestrus)		3.2	44.6	35.5	11.1	8.7

the pattern of uterine contractility (14). *Croker and Shelton* (15) have recently reported changes in the pattern of uterine contractility in spayed ewes following injection with differing doses of ODB. Their data are summarized in table VI.

With increasing doses of ODB there was an increase in the percentage of contractions moving towards the Fallopian tubes and a decrease in the number moving towards the body of the uterus. The pattern observed also differed from that observed in entire ewes.

Imbalance between levels of progesterone and oestrogen may also affect the qualitative and quantitative characteristics of cervical mucus, and the invasion of leucocytes from the uterine endometrium. Progesterone is a prerequisite for a cervical mucus response in spayed ewes ingesting oestrogenic forage (16), but *Allison* (17) has shown it not to be necessary for production of cervical mucus in spayed ewes following injections of ODB although there were significant differences in the pattern of response between ewes receiving or not receiving progesterone priming, respectively. These data are seen in figure 1. Differences occurred between 24 and 48 h after injection of ODB which was the period between insemination and slaughter in experiment 1. Dominance of either oestrogen or progesterone has thus been shown to effect both motility of the uterus (9, 15) and the production of cervical mucus (16, 17). These differences can be implicated in the differences in populations of spermatozoa recovered from the cervix following various hormonal regimes administered in the present work. Consistent with this concept is the fact that quite subtle changes in patterns of secretion of oestrogen and progesterone in progestagen-treated ewes (2) result in marked differences in the transport of spermatozoa in comparison with that observed in untreated ewes (3).

The comparative failure of sperm transport in spayed ewes as defined in experiment 2 cannot be explained on the same basis since major differences in sperm transport between progestagen-treated and untreated ewes are not evident

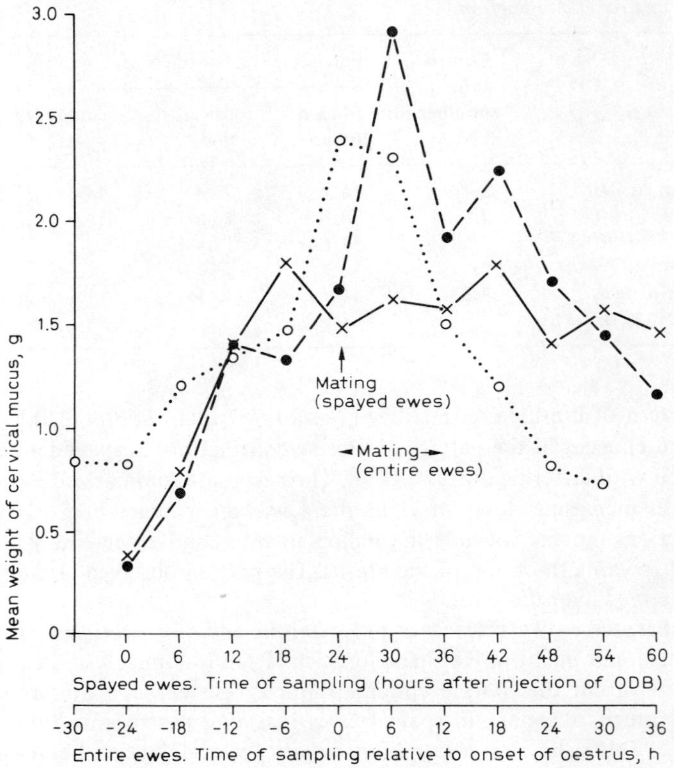

Fig. 1. Mean weights of cervical mucus produced at intervals of 6 h about the onset of oestrus in progestagen-treated entire ewes (·········) and in spayed ewes injected with 25 μg ODB (no progesterone priming = ——— ; 10 mg progesterone/day for 12 days = — — —). Data from *Allison* (17).

within 12 h of insemination (3). Although all ewes received identical treatment at insemination or mating the number of spermatozoa present in the cervix of spayed ewes 2 h after mating was only 31 % of the total recovered from entire animals. Further, a much greater proportion of that number were in the caudal section of the cervix.

The quantitative production of cervical mucus is comparable between entire ewes and spayed ewes in which oestrus has been induced (fig. 1) particularly within 2 h of insemination. It is unlikely that the characteristics of the mucus would affect survival and therefore motility of spermatozoa, as a high level of motility is maintained following incubation of spermatozoa for 6 h in mucus collected from the anterior vagina (18). Consequently, it is not easy to implicate differences in cervical secretions between the two groups in the reduced entry and subsequent progression of spermatozoa into the cervix in the spayed animals. In the 24 h following the onset of oestrus or mating in entire ewes there is a marked decline in production of cervical mucus (17) with coincident changes in molecular structure within that mucus. The physiological significance of this decline in production in relation to the maintenance of a population of spermatozoa within the cervix has not been determined.

There is a distinct possibility that the motility of the genital tract of the progesterone/oestrogen-treated spayed ewe differed from that of the normal oestrous ewe, so affecting the pattern of sperm transport and in particular the entry of spermatozoa into the cervix within 2 h of mating. Observed differences between uterine motility between spayed and entire ewes support this concept (15). There is no evidence concerning the leucocytic invasion from the uterine endometrium of spayed animals following injections of ovarian steroids; however, in progestagen-treated ewes an increased invasion of leucocytes at an earlier stage relative to the onset of oestrus in relation to that recorded in normal cyclic ewes (19) may be involved in the massive decline in numbers of spermatozoa between 12 and 24 h after insemination (3). The pattern of steroid release following intramuscular injections is unlikely to resemble the pattern of release in entire animals and certainly will not maintain the uterus in a normal physiological state. It is possible that this apparently inadequate endocrine regime may have contributed to the inefficient maintenance of the cervical population particularly in the cranial and mid sections, due to leucocytic invasion from the uterus.

Thus, differences in the transport of spermatozoa between spayed ewes subjected to different hormonal regimes and differences between spayed and entire ewes may be due to changing patterns of production of cervical mucus, genital tract motility and invasion of leucocytes from the uterus, although the relative contribution of each of these factors and the interactions between them are not clear.

In view of the differences in techniques used for recovery of spermatozoa from the cervix by different workers it is difficult to compare results. Techniques of simply flushing through the cervix although recovering approximately 90 % of the cervical mucus recover only about half of the total population of spermatozoa which is in contact with the majority of leucocytes within the cervix (20). If the cervix is opened longitudinally and then flushed, a higher proportion of spermatozoa may be recovered. Following flushing the cervix may be fixed, stained and sectioned, the total number of spermatozoa being estimated from counts of cross sections (20). This technique allows an estimation of total numbers within the lumen of the cervix plus a measure of those remaining following flushing and their distribution within the cervix. Finally, the technique used in experiment 2 recovers almost all spermatozoa in each section but does not differentiate between the population within the lumen, or closer to the mucosa and within the cervical crypts. In view of the suggested physiological significance of the spatial separation between spermatozoa and leucocytes in the cervix (20) and the maintenance of an adequate population after mating, a combination of the latter two techniques would seem more suitable for future research.

References

1 *Moore, N.W.; Quinlivan, T.D.; Robinson, T.J., and Smith, J.F.:* The fertilization of ova following natural mating or artificial insemination after withdrawal of SC-9880 impregnated intravaginal sponges; in The control of the ovarian cycle in the sheep, pp. 169–176 (Sydney University Press, Sydney 1967).

2 *Smith, J.F. and Robinson, T.J.:* The effect of exogenous progestagen on the ovarian vein plasma levels of free oestrogens in the Merino ewe. J. Endocrin. *48:* 485–496 (1970).

3 *Quinlivan, T.D. and Robinson, T.J.:* Numbers of spermatozoa in the genital tract after artificial insemination of progestagen-treated ewes. J. Reprod. Fertil. *19:* 73–86 (1969).

4 *Allison, A.J. and Robinson, T.J.:* The recovery of spermatozoa from the reproductive tract of the spayed ewe treated with progesterone and oestrogen. J. Reprod. Fertil. *31:* 215–224 (1972).

5 *Allison, A.J.:* A comparison of the transport of spermatozoa in spayed and entire ewes. J. Reprod. Fertil. *31:* 415–423 (1972).

6 *Mattner, P.E. and Braden, A.W.H.:* Spermatozoa in the genital tract of the ewe. I. Rapidity of transport. Austr. J. Biol. Sci. *16:* 473–481 (1963).

7 *Vickery, B.H. and Bennet, J.P.:* The cervix and its secretions in mammals. Physiol. Rev. *48:* 135–154 (1968).

8 *Reynolds, S.R.M.:* Physiology of the uterus (Hafner Publishing, Kingston-upon-Thames 1965).

9 *Brinsfield, T.H.:* Relationship of endocrine state to spontaneous motility of sheep myometrium *in vitro*. J. anim. Sci. *27:* 439–442 (1968).

10 *Odeblad, E.:* Types of human cervical secretion. Acta europ. fertil. *1:* 99–116 (1969).

11 *Lightfoot, R.J. and Restall, B.J.:* Effects of site of insemination, sperm motility and genital tract contractions on transport of spermatozoa in the ewe. J. Reprod. Fertil. *26:* 1–13 (1971).

12 *Brinsfield, T.H. and Hawk, H.W.:* Modifications of the direction of uterine contractions by intra-uterine devices in the ewe. J. Reprod. Fertil. *18:* 535–537 (1969).

13 *Hawk, H.W.:* Investigations into the antifertility effect of intra-uterine devices in the ewe. J. Reprod. Fertil. *14:* 49–59 (1967).

14 *Hawk, H.W. and Echternkamp, S.E.:* Uterine contractions in the ewe during progestagen-regulated oestrus. J. Reprod. Fertil. *34:* 347–349 (1973).

15 *Croker, K.P. and Shelton, J.N.:* Influence of stage of cycle, progestagen treatment and dose of oestrogen on uterine motility in the ewe. J. Reprod. Fertil. *32:* 521–524 (1973).

16 *Lindsay, D.R. and Francis, C.M.:* Cervical mucus measurement in ovariectomized ewes

as a bioassay of synthetic and phytooestrogens. Austr. J. agric. Res. *19:* 1069–1076 (1968).

17 *Allison, A.J.:* Production of cervical mucus in ewes treated with exogenous progestagen and oestrogen. Austr. J. agric. Res. *23:* 473–481 (1972).

18 *Restall, B.J.:* The influence of fluids from the genital tract of the ewe on the survival of ram spermatozoa. Austr. J. Biol. Sci. *22:* 1517–1525 (1969).

19 *Quinlivan, T.D.:* Studies in ovine reproduction; Ph.D. Thesis, Sydney (1967).

20 *Mattner, P.E.:* The distribution of spermatozoa and leucocytes in the female genital tract in goats and cattle. J. Reprod. Fertil. *17:* 253–261 (1968).

Author's address: Dr. *A.J. Allison,* Ministry of Agriculture and Fisheries, Invermay Agricultural Research Center, Private Bag, *Mosgiel* (New Zealand)

The Biology of Spermatozoa
INSERM Int. Symp., Nouzilly 1973, pp. 224–231 (Karger, Basel 1975)

Factors Affecting Sperm Transport and Uterine Motility in the Estrous Ewe

H. W. Hawk

Animal Physiology and Genetics Institute, USDA, Agricultural Research Center, Beltsville, Md.

A number of investigations have been conducted in recent years on sperm survival and transport in the ewe. Several of these studies were concerned with sperm transport following either the insertion of an intrauterine device into the ewe or the use of exogenous progestagen to regulate the estrous cycle (3, 10). This report is concerned primarily with the results of some of the experiments conducted in this laboratory.

Factors Affecting Ovum Fertilization and Sperm Transport

Materials and Methods

Animals

The ewes were predominantly Rambouillets, 6–8 years of age. They were checked for estrus twice daily by the use of vasectomized rams. The first day of estrus was designated as day 0 of the estrous cycle.

Ovum Fertilization

Ewes were mated naturally during estrus and killed 3 days later. The oviducts were flushed with saline solution and the ova were examined for cleavage and for the presence of accessory spermatozoa in the zona pellucida (3).

Recovery of Sperm Cells from the Reproductive Tract

Ewes were mated naturally during early estrus and killed 4–24 h after mating. Each segment of the reproductive tract was flushed separately with saline solution and sperm cells in the flushings were counted in a hemocytometer under a phase contrast microscope (4).

Intrauterine Devices

Each intrauterine device (IUD) was made from a strip or rod of polyethylene plastic that was formed into a cylindrical spiral shape (3). A spiral IUD was generally about 5 cm long; the outside diameter was 6–8 mm, and the turns of the spiral were about 1 cm apart.

An IUD was inserted by exposing the uterus through a mid-ventral laparotomy, puncturing the uterine wall with sharp-pointed scissors, and screwing the spiral IUD into the uterine lumen.

Progestagens

Progesterone in corn oil was administered by subcutaneous injection at 10 or 20 mg per day. Medroxyprogesterone acetate (MAP; The Upjohn Company) was administered orally at 60 mg per day in a soybean meal mix in a gelatin capsule; MAP was also administered by intravaginal sponge, each sponge containing 60 mg of the progestagen. Melengestrol acetate (MGA; The Upjohn Company) was administered orally at 0.3 mg per day in soybean meal in a gelatin capsule.

Progesterone and the synthetic progestagens given orally were administered once daily for about 17 days, beginning on day 8 of an estrous cycle. The intravaginal sponge was inserted on day 8 and left in place for 17 days.

Prostaglandin

The prostaglandin was $F_{2\alpha}$-THAM salt (The Upjohn Company). It was dissolved in water and administered by intramuscular injection.

Results and Discussion

Inhibition of Ovum Fertilization by an IUD

The presence of an IUD in the ewe for 2–4 weeks or for 8–10 months prevented ovum fertilization (3, 4). Cleaving ova were recovered from control ewes but not from IUD-bearing ewes (table I). An IUD in one uterine horn inhibited ovum fertilization even when ovulation occurred from the ovary opposite to the IUD-bearing horn; these results indicated that fertilization failure was due to something more than physical interference by the IUD with sperm transport through the uterus.

In a subsequent experiment, an IUD was inserted into ewes within a few hours after the beginning of estrus, then the ewes were mated (5). Ova were fertilized in control ewes but not in IUD-bearing ewes, indicating that ovum fertilization could be inhibited rapidly by the insertion of an IUD.

Inhibition of Sperm Transport by an IUD

The presence of an IUD in one uterine horn almost totally prevented sperm transport to both oviducts (table II). The lack of sperm cells in the oviducts was particularly apparent in the ewes killed 16–24 h after mating, a period covering the time of ovulation. Sperm transport into the uterine horns was also reduced (table II). In the oviducts and uterine horns of IUD-bearing ewes in which sperm cells were found, the numbers were only a very small percentage of the numbers found in control ewes (4).

The effect of an IUD in inhibiting sperm transport was found in ewes bearing an IUD for a few hours (5), for 2–4 weeks (4), for 8–10 months (4), or

Table I. Inhibition of ovum fertilization by intrauterine plastic spirals (3)

Ewe group	Number of ewes	Ewes ovulating	Ovulation points	Ova recovered	Ova cleaved	Ova with accessory sperm cells
Control	11	11	12	11	11	11
IUD	16	16	18	16	0	0

Table II. Effect of an IUD on sperm cell recovery from the oviducts and uterine horns (4)

Ewe group	Hours after mating	Number of ewes	Proportion of individual uterine horns or oviducts containing spermatozoa	
			horns	oviducts
Control	4–12	6	10/12	8/12
IUD	4–12	6	7/12	2/12
Control	16–24	7	13/14	14/14
IUD	16–24	7	2/14	0/14

for 2 years (14). In ewes with an IUD in one uterine horn, sperm transport was consistently inhibited on both sides of the reproductive tract, indicating that the IUD disrupted the physiological mechanisms of sperm transport throughout the uterus.

Progestagen Treatment and Sperm Transport

The administration of progestagen to ewes for about 2 weeks results in a reduced ovum fertilization rate in ewes inseminated at the ensuing estrus (11). The lowered fertilization rate is attributable to reduced efficiency of sperm transport in the female reproductive tract (7, 10). The effects of progestagen in reducing the efficiency of sperm transport were not as severe as the effects of an IUD.

Prostaglandin Treatment and Sperm Transport

The administration of prostaglandin $F_{2\alpha}$ to ewes during the luteal phase of the estrous cycle causes the premature regression of corpora lutea; estrus occurs 2 or 3 days after treatment (6). When ewes were mated at the induced estrus and killed 24 h later, the number of sperm cells in the oviducts was significantly lower in the prostaglandin-treated ewes than in controls (6).

Factors Affecting Uterine Motility

Contractions of the uterus are generally considered to aid in transporting sperm cells from the cervix to the oviducts. Because of the drastic inhibition of sperm transport in IUD-bearing ewes, the possibility was considered that the IUD might suppress uterine motility. Various compounds that stimulate uterine motility were administered to IUD-bearing ewes, but there was no increase in the rate of ovum fertilization (13). Recordings were made of the motility of longitudinal strips of myometrium, but the amplitude of contractions was actually greater for myometrium from IUD-bearing uterine horns than from uteri of control ewes (1). An experiment was then conducted to determine the effect of an IUD on the direction of uterine contractions. Other experiments followed on the effects of progestagen or prostaglandin on uterine contractions.

Materials and Methods

In the first experiment (table III), uteri of estrous ewes were exposed by mid-ventral laparotomy, the motility of the uterus was recorded on movie film, and the number and direction of uterine contractions were obtained by viewing the developed film (2).

In subsequent experiments, the uterus was exposed and observed for 10 min. Individual contractions of the uterus were recorded as originating in the posterior, middle, or anterior segments of the uterine horns, and then moving toward the oviduct, toward the cervix, or not moving clearly in either direction. The contractions of the uterine horn recorded as moving were visible as constrictions that moved lengthwise along the horn in one direction or the other. With few exceptions, the contractions moving toward the oviducts originated in the posterior segments of the uterine horns. The contractions moving toward the cervix originated near the tubo-uterine junction. The contractions with 'indefinite direction' (tables III–VI) did not have a clearly discernible directional movement. This classification includes powerful contractions that uncoiled and straightened the entire horn, contractions that covered a length of several centimeters in one or more segments of the horn but did not clearly move lengthwise, localized constrictions that did not move lengthwise and, in some instances, weak contractions of the myometrium indicated by a blanching of the serosa.

Results and Discussion

Intrauterine Devices

The normal pattern of uterine motility was drastically disrupted during estrus in IUD-bearing ewes (2). The number of contractions moving toward the oviducts was greatly decreased and the number moving toward the cervix was increased (table III). Confirmation of the change in direction of uterine contractions in IUD-bearing ewes was provided by recordings of uterine contractions *in vivo* through open-ended catheters placed in the uterine lumen (9).

Table III. Effect of an IUD on uterine contractions during estrus (2)

Ewe group	Number of ewes	Number of contractions per ewe per 10 min[1]			
		total	moving toward oviducts	indefinite direction	moving toward cervix
Control	8	69	40 (58)	18 (26)	11 (16)
IUD	8	55	9 (17)	7 (13)	39 (70)

1 Figures in parentheses are percentages of the total number of contractions.

Table IV. Uterine contractions during progestagen-regulated estrus (8)

Ewe group[2]	Number of ewes	Number of contractions per ewe per 10 min[1]			
		total	moving toward oviducts	indefinite direction	moving toward cervix
Controls	20	63	34 (54)	21 (33)	8 (13)
Progestagen	53	52	11 (22)	21 (40)	20 (38)

1 Figures in parentheses are percentages of the total number of contractions.
2 In control ewes, 92 % of all contractions moving toward the oviducts originated in the posterior segments of the uterine horns; 94 % of the contractions moving toward the cervix originated in the anterior ends of the horns. In progestagen-treated ewes, the proportions were 95 and 96 %, respectively.

Table V. Effect of prostaglandin $F_{2\alpha}$-induced regression of corpora lutea on uterine contractions during the ensuing estrus (6)

Treatment	Number of ewes	Number of contractions per ewe per 10 min[1]			
		total	moving toward oviducts	indefinite direction	moving toward cervix
None	11	56	30 (54)	17 (30)	9 (16)
PG $F_{2\alpha}$	15	43	18 (42)	16 (37)	9 (21)
MAP	12	51	11 (22)	23 (45)	17 (33)

1 Figures in parentheses are percentages of the total number of contractions.

An IUD affected the direction of uterine contractions in estrous ewes as early as a few hours or as long as 2 years after the IUD was inserted (5, 14). The consistent 'reversal' in the direction of uterine contractions in IUD-bearing ewes may explain the inhibition of sperm transport (2, 5, 14).

Progestagens

Ewes were treated with MAP orally or intravaginally, with MGA orally, or with progesterone subcutaneously (8). Uterine contractions were observed when the ewes were in estrus, 2–4 days after the end of progestagen treatment. None of the ewes had ovulated at the time contractions were observed.

The prior treatment of ewes with progestagen significantly decreased the number of contractions moving toward the oviducts, increased the number moving toward the cervix, and tended to reduce the total number (table IV). The altered uterine contractions may account at least in part for the decreased efficiency of sperm transport to the oviducts at the estrus following progestagen treatment (7, 10).

The footnote to table IV shows the relationship between the site of origin of uterine contractions along the uterine horn and the direction of movement of the contractions.

Prostaglandin

Because of the potential usefulness of prostaglandin for controlling estrus and ovulation, an experiment was conducted to determine the effect of prostaglandin treatment during the luteal phase of the estrous cycle on uterine motility at the induced estrus (6). Ewes treated intravaginally with MAP were included for comparison. The number of uterine contractions moving toward the oviducts in prostaglandin-treated ewes was significantly lower than in controls, but tended to be higher than in MAP-treated ewes (table V). Prostaglandin, unlike MAP, did not affect the number of contractions moving toward the cervix.

The effect of prostaglandin treatment on uterine contractions during the ensuing estrus might explain the reduced efficiency of sperm transport to the oviducts during prostaglandin-induced estrous periods (6).

Stage of Estrus

Ewes were checked continually for estrus during daylight hours, and only ewes that came into estrus during daylight were used. Laparotomy was performed at either 4–6, 24–28 or 46–50 h after the beginning of estrus (table VI). Each ewe was examined at only one time interval. At 4–6 h, no ewes had ovulated; at 24–28 h, 9 of 10 ewes had ovulated but all were still in estrus; at 46–50 h, all ewes had ovulated and none were in estrus.

The number and origin of contractions at 4–6 h (table VI) were similar to those observed before ovulation in control ewes of other experiments (tables

Table VI. Effect of time after the beginning of estrus on the direction of uterine contractions

Hours after beginning of estrus	Number of ewes	Number of contractions per ewe per 10 min[1]			
		total	moving toward oviducts	indefinite direction	moving toward cervix
4–6	10	57	37 (65)	14 (25)	6 (10)
24–28	10	48	20 (42)	12 (25)	16 (33)
46–50	10	46	6 (13)	4 (9)	36 (78)

1 Figures in parentheses are percentages of the total number of contractions.

III–V). The number of contractions originating in the posterior segments of the uterus and moving toward the oviducts had decreased significantly by 24–28 and 46–50 h. The number of contractions originating near the tubo-uterine junctions and moving toward the cervix had increased significantly at both 24–28 and 46–50 h. The results demonstrated clearly that the nature of uterine contractions changed between early estrus (4–6 h) and late estrus (24–28 h).

The appearance of uterine contractions also changed over the time periods studied. During early estrus, the contractions were strong and moved rapidly. By 46–50 h, most of the contractions began in the anterior ends of the uterine horns as deep constrictions that moved slowly in a posterior direction.

Recent results of experiments on uterine contractions have shown that exogenous estradiol administered to ovariectomized ewes initiates uterine contractions that originate in the posterior segments of the uterus and move toward the oviducts. Estradiol given to ewes in estrus increased the number of contractions originating in the posterior segments of the uterus and maintained this type of contraction until at least 48 h after the beginning of estrus. When the ovaries were removed from estrous ewes before ovulation, the origin of a majority of contractions changed by 48 h, as in intact ewes, from the posterior to the anterior segments of the uterine horns; exogenous estradiol prevented the change. These results suggest that the decline in endogenous estrogen secretion that begins soon after the start of estrus (12) causes the change in direction of contractions that takes place between early estrus and 48 h after the beginning of estrus.

General Discussion

These experiments have shown that both sperm transport and uterine contractility in the estrous ewe can be changed by the presence of an IUD or by prior treatment of the animal with progestagen or prostaglandin. Also, the type of contractions begins to change naturally before the end of estrus. Endogenous estrogen very likely exerts considerable control over the origin and direction of uterine contractions during estrus, and declining estrogen secretion probably accounts for the change in the nature of contractions before the end of estrus.

The uterine motility pattern that appears to be 'normal' in the estrous ewe is both transitory and susceptible to change by a number of experimentally imposed factors. The physiological pathways that control both the normal and induced changes in motility patterns need to be elucidated.

References

1 *Brinsfield, T.H. and Hawk, H.W.:* Stimulation by intra-uterine devices of motility of sheep myometrium. J. Reprod. Fertil. *16:* 129–131 (1968).

2 *Brinsfield, T.H. and Hawk, H.W.:* Modification in the direction of uterine contractions by intra-uterine devices in the ewe. J. Reprod. Fertil. *18:* 535–537 (1969).

3 *Hawk, H.W.:* Inhibition of ovum fertilization in the ewe by intrauterine plastic spirals. J. Reprod. Fertil. *10:* 267–269 (1965).

4 *Hawk, H.W.:* Investigations into the anti-fertility effect of intrauterine devices in the ewe. J. Reprod. Fertil. *14:* 49–59 (1967).

5 *Hawk, H.W.:* Rapid disruption of sperm transport mechanisms by intrauterine devices in the ewe. J. Reprod. Fertil. *23:* 139–142 (1970).

6 *Hawk, H.W.:* Uterine motility and sperm transport in the estrous ewe after prostaglandin induced regression of corpora lutea. J. anim. Sci. *37:* 1380–1385 (1973).

7 *Hawk, H.W. and Conley, H.H.:* Investigation of sperm transport failures in ewes administered synthetic progestagen. J. anim. Sci. *34:* 609–613 (1972).

8 *Hawk, H.W. and Echternkamp, S.E.:* Uterine contractions in the ewe during progestagen-regulated estrus. J. Reprod. Fertil. *34:* 347–349 (1973).

9 *Mann, L.I.:* Effect of the IUCD on uterine motility in the ewe. Fertil. Steril. *20:* 951–958 (1969).

10 *Quinlivan, T.D. and Robinson, T.J.:* Numbers of spermatozoa in the genital tract after artificial insemination of progestagen-treated ewes. J. Reprod. Fertil. *19:* 73–86 (1969).

11 *Robinson, T.J.:* The control of the ovarian cycle in the sheep, p. 169 (Sydney University Press, Sydney 1967).

12 *Smith, J.F. and Robinson, T.J.:* The effect of exogenous pregestagen on the levels of free oestrogen in the ovarian vein plasma of the ewe. J. Endocrin. *48:* 485–496 (1970).

13 *Warren, J.E., jr. and Hawk, H.W.:* Attempts to overcome the antifertility effect of intrauterine plastic spirals in the ewe. J. anim. Sci. *27:* 131–133 (1968).

14 *Warren, J.E., jr. and Hawk, H.W.:* Effect of an intrauterine device on sperm transport and uterine motility in sheep and rabbits. J. Reprod. Fertil. *26:* 419–422 (1971).

Author's address: *H.W. Hawk,* Reproduction Laboratory, Animal Physiology and Genetics Institute, Agricultural Research Center, Building 161, Barc-East, *Beltsville, MD 20705* (USA)

The Biology of Spermatozoa
INSERM Int. Symp., Nouzilly 1973, pp. 232–238 (Karger, Basel 1975)

Fertility and Sperm Transport following Prostaglandin or Progestogen Treatment of Cattle

J.W. Lauderdale

Agricultural Research and Development, The Upjohn Company, Kalamazoo, Mich.

I. Introduction

Progestogens administered for 14 or more days have been shown to be effective as means of regulating estrus in cattle and sheep. However, breeding at the synchronized estrus following progestogen withdrawal has resulted in reduced fertility in cattle.

The report of *Pharriss and Wyngarden* (14) that prostaglandin $F_{2\alpha}$ was luteolytic in the pseudopregnant rat was followed by reports that $PGF_{2\alpha}$ was luteolytic in cattle (6, 7, 10, 11, 15) and sheep (1–3, 12, 13, 16). Cattle returned to estrus within about three days after intrauterine, intramuscular, or subcutaneous administration of $PGF_{2\alpha}$ if injections were made after but not before 5 days after estrus. Thus, these data supported that $PGF_{2\alpha}$ could be utilized effectively to control estrus if administered to cattle which had a corpus luteum (CL).

The primary objective of this paper is to present data on pregnancy rates of cattle inseminated at estrus following prostaglandin administration. This objective seems warranted since reduced fertility was a major detriment related to synchronization of estrus of cattle by means of progestogens. For comparative purposes, published data were summarized and are presented for pregnancy rates of cattle inseminated at estrus following progestogen treatment. No attempt will be made to document effectiveness of prostaglandin administration schemes which might be employed to control estrus of cattle maintained under routine herd management programs.

II. Fertilization Rates and Pregnancy Rates following Progestogen Treatment

The progestogens, doses, and routes of administration discussed are summarized in table I.

Fertilization rates. Ova were recovered at either 48 h after ovulation or 84 h after start of estrus. Fertilization rates were 66 and 49 % for control and DHPA-fed cattle (table II) (21). Fertilization rates were 82, 61, and 82 %, for ova recovered 2.5–3.0 days after insemination for heifers of control, CAP and AI at first estrus, and CAP and AI at second estrus, respectively (table II) (18).

Pregnancy rates (percentage of cattle inseminated that were pregnant about 35–60 days after insemination). First service pregnancy rates were (table III): (1) 58 and 38 % for control and MAP fed heifers (17); (2) 67 and 49 % for control and CAP fed heifers (18); (3) 55 and 35 % for control and MGA® fed heifers (8); and (4) 61 and 57 % for control and DHPA fed plus estradiol valerate treated heifers (20).

These data demonstrate fertility was reduced, as measured by either fertilization rates or pregnancy rates, for insemination made at first synchronized estrus following feeding of various progestogens (fig. 1). However, data on the short interval (9 days) of progestogen feeding suggests no significant alteration in pregnancy rate (20).

Table I. Summary of progestogens, doses, and routes of administration

Compound	Dose administered			Return to estrus (days)	Reference
	per day (mg)	number of days	route		
Dihydroxyprogesterone acetophenide (DHPA)	500	20	oral	2–4	(21)
DPHA plus estradiol valerate	400 5	9 1[1]	oral intramuscular	2–4	(20, 21)
Chlormadinone acetate (CAP)	10, 5[2]	14, 4[2]	oral	2–7	(18)
Medroxyprogesterone acetate (MAP)	180	18	oral	2–6	(17)
Melengestrol acetate (MGA®)	0.5 or 1.0	14	oral	3–8	(8)

1 Injected on day 2 of DHPA administration.
2 Initial 14 days at 10 mg, subsequent 4 days at 5 mg.

Table II. Fertility after progestogens

Treatment	Pregnancy rate			Treatment	Fertilization rate		
	number of cattle	first service pregnancy rate, %[1]	reference		number of ova recovered	percent ova fertilized	reference
Control	5,196	58	(17)	Control	32	66	(21)
MAP	2,337	38		DHPA	37	49	
Control	487	67	(18)	Control	45	82	(18)
CAP	413	49		First estrus after CAP	44	61	
Control	2,914	55	(8)	Second estrus after CAP	11	82	
MGA	531	35					
Control	1,156	61	(21)				
DHPA implant + estradiol valerate	1,227	57					

1 Pregnancy rate = number pregnant/number AI.

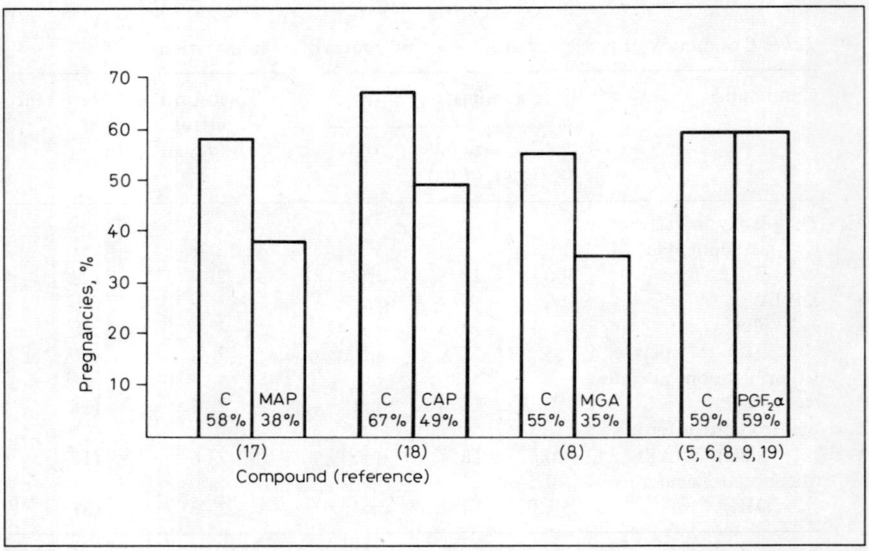

Fig. 1. Pregnancy following insemination at first estrus detected after either prostaglandin or oral progestogen administration.

Table III. Fertility of cows inseminated after prostaglandin treatment

Treatment	Dose and route of administration	Duration of breeding[1]	Number of animals	Percent pregnant[2]	Reference
Control	none	21 days	12	58	(8)
$PGF_{2\alpha}$	30 mg subcutaneous[3]	days 2–5 after $PGF_{2\alpha}$	17	59	
Control	none	18–25 days	122	53	
$PGF_{2\alpha}$	30 mg subcutaneous or intramuscular[3]	days 1–7 after $PGF_{2\alpha}$	69	52	(9)
$PGF_{2\alpha}$	30 mg subcutaneous or intramuscular[3]	72 and 90 h after $PGF_{2\alpha}$	86	56	
Control	none	27 days	20	65	(5)
Progestogen implant + $PGF_{2\alpha}$	7-day implant 30 mg $PGF_{2\alpha}$ intramuscular at implant removal[3]	days 2–5 after $PGF_{2\alpha}$	47	64	
Control	none	15 days	26	62	(6)
$PGF_{2\alpha}$	1 mg intrauterine[4]	days 2.5–3.3 after $PGF_{2\alpha}$	17	59	
$PGF_{2\alpha}$	1.5 or 2.0 mg intrauterine	days 2.5–3.3 after $PGF_{2\alpha}$	17	53	(6)
Control	none	7 days	19	79	(19)
$PGF_{2\alpha}$	2 mg intrauterine[4]	days 2.5–3.0 after $PGF_{2\alpha}$	22	77	
Control	none	21 days	17	65	(19)
$PGF_{2\alpha}$	5 mg intrauterine	days 2.5–6.0 after $PGF_{2\alpha}$	17	53	
All controls			216	59	
$PGF_{2\alpha}$ AI at estrus[5]			189	59	
All $PGF_{2\alpha}$			292	58	

1 Artificial insemination associated with twice daily observations of cows for estrus.
2 Based or rectal palpation of the uterus 35–65 days after insemination.
3 Dose was $PGF_{2\alpha}$-THAM salt.
4 400 μg estradiol benzoate administered intramuscular to some of the cattle.
5 Includes only studies with control cows.

III. Sperm Transport, Fertilization Rates, and Pregnancy Rates following Prostaglandin Treatment

Sperm transport and fertilization rates. No data were available on either sperm transport or fertilization rates for cattle, but such data has been reported for the ewe (3, 4). Number of intact spermatozoa recovered 24 h after mating from the vagina, cervix, or uterus did not differ significantly for control ewes, ewes in estrus after intramuscular injection of $PGF_{2\alpha}$-THAM salt on day 9 of an estrous cycle, or ewes in estrus after removal of 60 mg MAP intravaginal sponge, which had been in place for 16 days (4). Significantly ($p < 0.02$) fewer intact spermatozoa were recovered from the oviduct of prostaglandin treated (5.1×10^3) ewes as compared to control (20.7×10^3) ewes (4). In addition, although not statistically significant, fewer ewes had cleaving ova recovered for prostaglandin (7 of 10) as compared to either control (9 of 10) or MAP (10 of 10) treatments (4). Fewer control ewes (3 of 10) as compared to either prostaglandin (5 of 10) or MAP (7 of 10) ewes yielded ova with either no or 1–5 accessory sperm cells. These data (3, 4), based on the ewe, would suggest that pregnancy rates might be reduced following insemination made at estrus following $PGF_{2\alpha}$ treatment.

Pregnancy rates. Dose of $PGF_{2\alpha}$, routes of administration, duration of breeding, and number of animals are presented in table III for several studies executed to investigate pregnancy rate of cattle inseminated after prostaglandin treatment. Pregnancy was determined by palpation of the uterus per rectum between about 35 and 65 days after artificial insemination. Pregnancy rates were 58 and 59 % for control and $PGF_{2\alpha}$-treated cattle if $PGF_{2\alpha}$ was administered between days 6 and 16 of the estrous cycle (table III) (8).

In a subsequent study, designed to investigate pregnancy rates of $PGF_{2\alpha}$-treated cattle with differing genetic bases and managed under differing environmental conditions, pregnancy rate comparisons were made among the following treatment groups: (1) controls inseminated at estrus detected over an 18- to 25-day interval; (2) $PGF_{2\alpha}$ injected and inseminated at estrus detected within 7 days after $PGF_{2\alpha}$, and (3) $PGF_{2\alpha}$ injected, responded within 7 days after injection, but inseminated only at about 72 and 90 h after $PGF_{2\alpha}$ (9). Pregnancy rates were 53, 52, and 56 %, respectively (table III).

Pregnancy rates were compared between controls and heifers implanted for 7 days with a progestogen and injected with $PGF_{2\alpha}$ on the day of implant removal (5). Pregnancy rates were 65 and 64 % (table III).

Pregnancy rates following intrauterine $PGF_{2\alpha}$ administration, for a series of studies completed over the past two years, have been (table III): 62 and 59 % for control and $PGF_{2\alpha}$ (6); 53 % for $PGF_{2\alpha}$ (6); 79 and 77 % for control and $PGF_{2\alpha}$ (19); and 65 and 53 % for control and $PGF_{2\alpha}$ (19).

If one pools the data within control and within $PGF_{2\alpha}$-treated cattle inseminated at estrus within 7 days after injection for the six studies cited above, pregnancy rates were 59 and 59 %, respectively (fig. 1). Thus, pregnancy rate of cattle inseminated during synchronized estrus following $PGF_{2\alpha}$ administration was not different from that for control cattle.

IV. Distribution of Estrus after $PGF_{2\alpha}$

The percent of cattle detected in estrus on days 1, 2, 3, 4, 5, 6, and 7 after $PGF_{2\alpha}$ administration was 2, 20, 63, 10, 3, 1, and 1, respectively. The data do not reflect number of cattle treated with prostaglandin but not detected in estrus during the subsequent 7 days. Future research should be conducted to investigate various management schemes which could be employed so that 100 % of a group of cattle could be either detected in estrus or inseminated within a predefined synchronized interval. Two such schemes which have been investigated are: (1) two injections of prostaglandins administered to all cattle at an interval of 10–12 days of each other, and (2) administration of prostaglandins on the last day of a 5- to 7-day progestogen treatment.

References

1 *Barrett, S.; Blockey, M.A. de; Brown, J.M.; Cumming, I.A.; Goding, J.R.; Mole, B.J., and Obst, J.M.:* Initiation of the oestrus cycle in the ewe by infusions of $PGF_{2\alpha}$ to the autotransplanted ovary. J. Reprod. Fertil. *24:* 136–137 (1971).

2 *Douglas, R.H. and Ginther, O.J.:* Effect of prostaglandin $F_{2\alpha}$ in ewes and pony mares. J. anim. Sci. *37:* 308 (1973).

3 *Hawk, H.W.:* Estrus, uterine motility, and sperm transport after prostaglandin. J. anim. Sci. *37:* 314 (1973).

4 *Hawk, H.W.:* Uterine motility and sperm transport in the estrous ewe after prostaglandin-induced regression of corpora lutea. J. anim. Sci. *37:* 1380–1385 (1973).

5 *Heersche, B.G.; Kiracofe, G.H.; McKee, R.M.; Davis, D.L., and Brower, G.R.:* Control of estrus in heifers with $PGF_{2\alpha}$ and Syncromate B. J. anim. Sci. *38:* 225 (1974).

6 *Inskeep, E.K.:* Potential uses of prostaglandins in control of reproductive cycles of domestic animals. J. anim. Sci. *36:* 1149–1157 (1973).

7 *Lauderdale, J.W.:* Effects of $PGF_{2\alpha}$ on pregnancy and estrous cycle of cattle. J. anim. Sci. *35:* 246 (1972).

8 *Lauderdale, J.W.:* Fertility and sperm transport following prostaglandin or progestogen treatment of cattle; in Transport, survie et pouvoir fécondant des spermatozoïdes chez les vertébrés (INSERM, Paris 1973).

9 *Lauderdale, J.W.; Seguin, B.E.; Stellflug, J.N.; Chenault, J.R.; Thatcher, W.W.; Vincent, C.K., and Loyancano, A.F.:* J. anim. Sci. *38:* (in press, 1974).

10 *Liehr, R.A.; Marion, G.B., and Olson, H.H.:* Effects of prostaglandin on cattle estrous cycles. J. anim. Sci. *35:* 247 (1972).

11 *Louis, T.M.; Hafs, H.D., and Morrow, D.A.:* Estrus and ovulation after uterine $PGF_{2\alpha}$ in cows. J. anim. Sci. *35:* 247–248 (1972).

12 *McCracken, J.A.; Baird, D.T., and Goding, J.R.:* Factors affecting the secretion of steroids from the transplanted ovary in the sheep. Recent Progr. Hormone Res. *27:* 537–582 (1971).

13 *McCracken, J.A.; Carlson, J.C.; Glew, M.C.; Goding, J.R.; Baird, D.T.; Green, K., and Samuelson, B.:* Prostaglandin $F_{2\alpha}$ identified as a luteolytic hormone in the sheep. Nature new Biol. *238:* 129–134 (1972).

14 *Pharriss, B.B. and Wyngarden, L.J.:* The effect of $PGF_{2\alpha}$ on the progesterone content of ovaries from pseudopregnant rats. Proc. Soc. exp. Biol. Med. *130:* 92–94 (1969).

15 *Rowson, L.E.A.; Tervit, R., and Brand, A.:* The use of prostaglandin for synchronization of oestrus in cattle. J. Reprod. Fertil. *29:* 145 (1972).

16 *Thorburn, G.D. and Nicol, D.H.:* Regression of the ovine corpus luteum after infusion of prostaglandin $F_{2\alpha}$ into the ovarian artery and uterine vein. J. Endocrin. *51:* 785–786 (1971).

17 Upjohn Company, The: The Repromix® story, pp. 10–19 (1965).

18 *Wagner, J.F.; Veenhuizen, E.L.; Gregory, R.P., and Tonkinson, L.V.:* Fertility in the beef heifer following treatment with 6-chloro-Δ^{6-17}-acetoxyprogesterone. J. anim. Sci. *27:* 1627–1630 (1968).

19 *Welch, J.A. and Inskeep, E.K.:* Unpublished data (1973).

20 *Wiltbank, J.N.:* Short breeding periods with the aid of estrous cycle synchronization. Proc. 3rd Tech. Conf. on Artificial Insemination and Reproduction. Nat. Ass. Animal Breeders, Chicago 1970, pp. 85–90.

21 *Wiltbank, J.N.; Shumway, R.P.; Parker, W.R., and Zimmerman, D.R.:* Duration of estrus, time of ovulation, and fertilization rate in beef heifers synchronized with dihydroxyprogesterone acetophenide. J. anim. Sci. *26:* 764–767 (1967).

Author's address: Dr. *J.W. Lauderdale,* Agricultural Division, The Upjohn Company, *Kalamazoo, MI 49001* (USA)

The Biology of Spermatozoa
INSERM Int. Symp., Nouzilly 1973, pp. 239–249 (Karger, Basel 1975)

Influence of Hormonal and Intrauterine Contraceptives on Human Sperm Migration *in vivo* and *in vitro*

E. Kesserü

Department of Obstetrics and Gynecology, University Cayetano Heredia, Lima

The present study was undertaken as an attempt to determine the role played by the alterations of sperm migration in the mechanism of action of hormonal and intrauterine contraceptives. Various currently used contraceptive agents were tested under both *in vivo* and *in vitro* conditions, by means of standardized techniques of sperm migration measurement (3).

Material

The subjects were young, healthy, fertile and eumenorrheic women attending our contraception and infertility clinics. The patients were distributed into groups, according to the contraceptive treatments, besides of a control group. Table I shows the distribution of the human material and the numbers of cases studied.

Control Group

These patients were selected with special care. Normal ovulatory cycles were present in all of them as determined by a biphasic basal body temperature, urinary pregnanediol, serial colpocytology and endometrial biopsies.

Groups with Hormonal Contraceptives

The group treated with *oral estrogen* consisted of women receiving sequential therapy; thus, during the midcycle days, they were taking pills containing estrogen alone (mestranol 0.08 mg daily). The *oral progestagen microdose* employed was *d*-norgestrel 0.03 mg/day. The *monthly injected* group received norethindrone enanthate, 25 mg, once a month. The *depo injected* group received norethindrone enanthate, 200 mg, which has been demonstrated to give contraceptive protection that extends for 12 weeks. This group was subdivided according to whether the test was performed during the first month after injection, the second month, or exactly 12 weeks after injection when the medication begins to loose its clinical effectiveness.

Table I. Groups of women studied

	In vivo sperm penetration tests	*In vitro* sperm migration tests
Control group	19	49
Hormonal contraceptives		
Oral estrogen	10	31
Progestogens		
Oral microdose	50	50
Monthly injection	30	21
Depo injection		
1st month postinjection	43	52
2nd month postinjection	18	26
12 weeks postinjection	32	34
Intrauterine contraceptives		
Nonmetallic		
Lippes' D and C	44	77
Aries	7	–
Metallic		
Aries-Cu	9	–
Aries-Cu/Ag and Cu/Ni	16	–
Total cases	278	340

Groups with Intrauterine Contraceptives

The *nonmetallic IUDs* included large sized Lippes' loops and a new design named 'Aries'. The *metallic IUDs* consisted of the Aries model employed as a vector to which metallic adjunts were attached. The metals used were copper alone (Aries-Cu) and copper together with silver or nickel (Aries-Cu/Ag and Aries-Cu/Ni).

Methods

In vivo *Sperm Penetration Test*

The methodology of our technique has been described elsewhere (2, 3). The postcoital test consisted of sperm countings in sequential specimens from cervical mucus and intrauterine fluid. For each specimen, three values were determined: (a) the total number of sperm; (b) the number of motile sperm, and (c) the number of sperm with directional motility, e.g. the number of sperm presenting progressive forward motion.

In vitro *Sperm Migration Test*

The capillary tube method described by *Kremer* (7) has been slightly modified (3, 6) in order to improve its clinical usefulness. The capillary tube was filled with the cervical mucus sample up to a length of 6 cm.

In each test, four parameters of sperm migration were studied: (a) penetration depth – the distance traveled by the most forward sperm, at 30 and 120 min from the time of initial mucus-semen contact; (b) penetration density – the number of sperm per LPF was determined at distances of 1, 2 and 5 cm, 120 min after initial contact; (c) qualitative motility – the degree of motility was registered at distances of 1, 2 and 5 cm at 120 min, using a conventional scale; (d) duration of motility – at 24 and 48 h after initial contact, the entire length of the capillary tube was examined in order to verify the presence of motile sperm.

Results

In vivo Sperm Penetration Test

With the aim of abbreviating the presentation, the results will refer only to the number of *motile sperm.* The results of the control group were the following (mean ± SEM): in the cervical canal specimen 16.5 ± 2.1; in the intrauterine fluid specimen 5.7 ± 0.8.

Groups with Hormonal Contraceptives

In cervical material (fig. 1), the value of the group on estrogen, with 34.4 ± 5.9, was significantly higher than the control ($p < 0.02$). Among the groups with progestagens, the only significantly lowered values, in comparison to the controls, was the depo-injected group at the first month postinjection: 6.6 ± 1.3 ($p < 0.01$). In the intrauterine specimen (fig. 2), the values of the estrogen group were similar to the controls ($p < 0.5$). Of great interest are the findings in the progestagen groups with an almost total or total *absence* of intrauterine sperm. Thus, positive sperm recoveries were obtained only in the oral microdose group (6 cases out of 50) and at 12 weeks postinjection of the depo injected group (7 of 32). Even in these few positive cases sperm were found only in very small number.

Groups with Intrauterine Contraceptives

In the intracervical specimen (fig. 3) no significant differences were observed between the various groups. In the intrauterine fluid (fig. 4) the values of the nonmetallic groups (Lippes and plain Aries) were similar to the control. On the contrary, in the groups with metal-bearing IUDs, intrauterine recoveries were negative in the great majority of cases.

In vitro Sperm Migration Test

Findings, grouped according to the four parameters of sperm migration evaluated, are shown in figures 5–8. The groups with hormonal contraceptives are the same as for the case of the *in vivo* results. Among the IUDs, in this study there was included only a group of women fitted with Lippes' loops (table I).

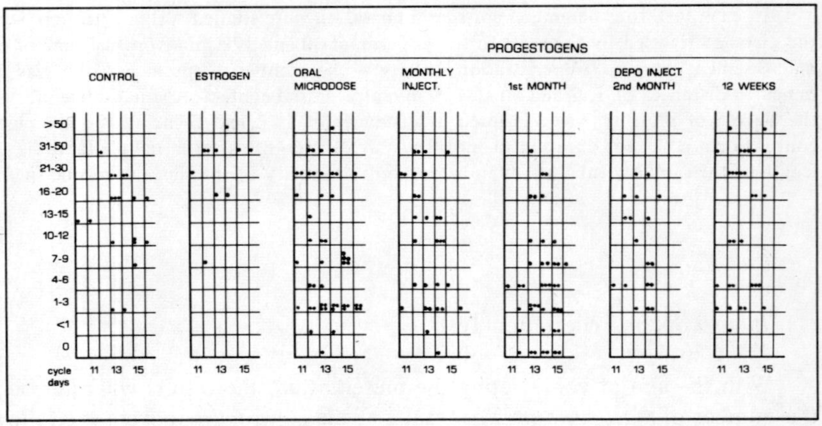

Fig. 1. Intracervical sperm recovery. Values of the number of motile sperm in the cervical mucus. Figures represent the average number of sperm per microscopic field (× 250). Results of the control group and the groups with hormonal contraceptives.

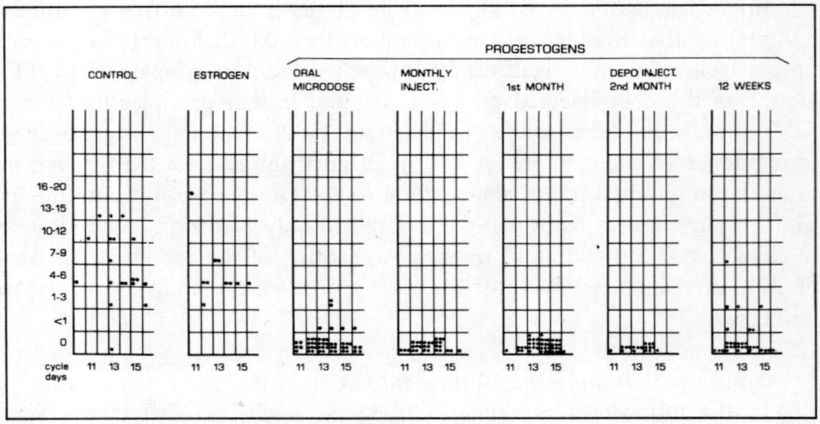

Fig. 2. Intrauterine sperm recovery. Values of the number of motile sperm in the uterine fluid. Rest of legends as in figure 1.

Penetration Depth (fig. 5)

The control group showed a penetration (mean ± SEM) of 4.85 ± 0.19 cm at 30 min and 5.69 ± 0.1 cm at 120 min. In the estrogen group, the penetration depth was significantly greater than the control group at 30 min as well as at 120 min ($p < 0.02$). In the IUD group the figures are dispersed over a wide range; however, the mean was not significantly different from the control group.

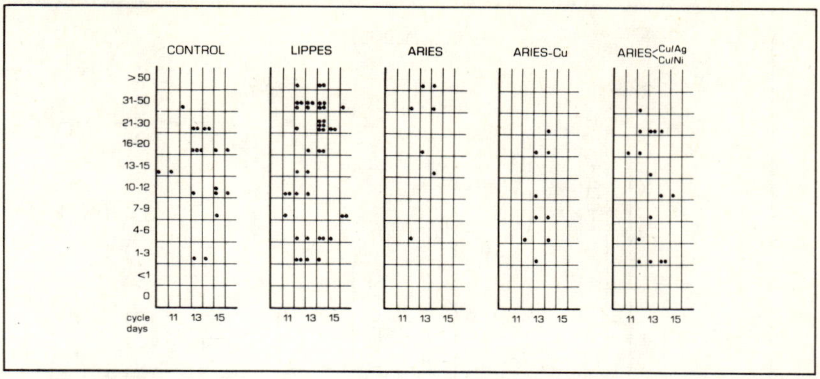

Fig. 3. Intracervical sperm recovery. Results of the control group and the groups with intrauterine contraceptives.

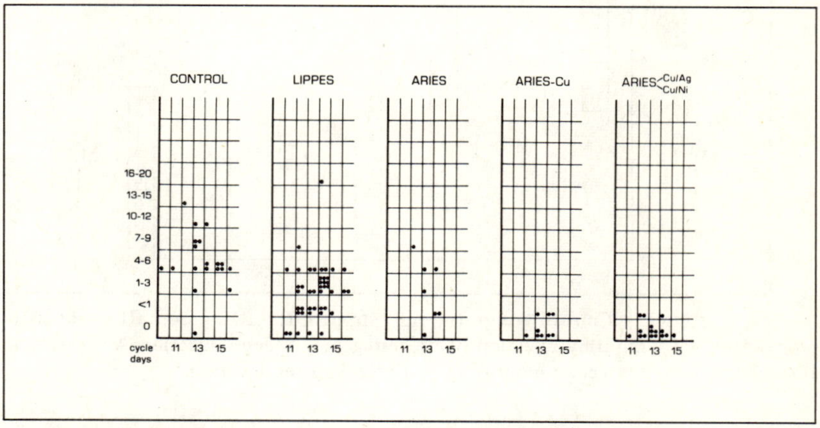

Fig. 4. Intrauterine sperm recovery. Rest of legends as in figure 3.

All of the progestagen-treated groups showed significantly decreased values ($p < 0.0001$). The means are inversely related to the progestagen dose. The following finding should be emphasized: in *no case* of the progestagen was penetration more than 3.5 cm at 30 min and 5 cm at 120 min, except for one patient of the oral microdose group that had a sperm penetration of 6 cm at 120 min.

Penetration Density (fig. 6)

The estrogen group showed a significant increase ($p < 0.001$) at all three distances. The IUD group was not significantly different than the control. The

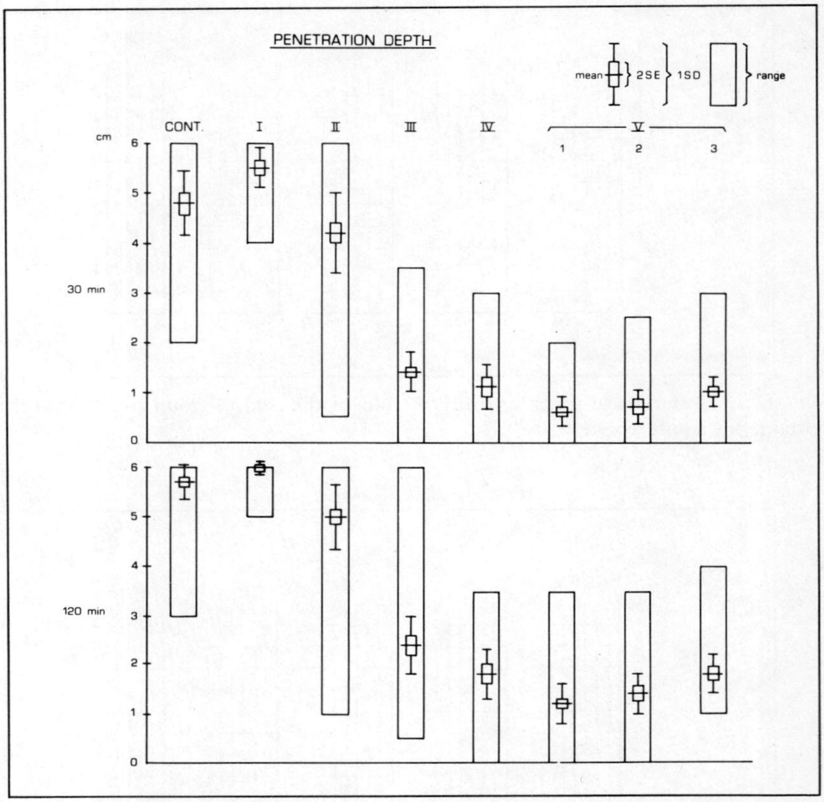

Fig. 5. CONT. = Control group; I = oral estrogen; II = IUD group; III = oral microdose progestagen; IV = monthly injection of progestagen; V = depo injection; V/1 = first month after injection; V/2 = second month; V/3 = 12 weeks after injection.

progestagen groups showed lower concentrations than the control group; however, at 1 cm distance, this difference is not very striking. This finding is in contrast to what was observed at 5 cm where there was not a single sperm found in any of the 183 patients treated with progestagens, save one case with the microdose, mentioned above.

Qualitative Motility (fig. 7)

With this parameter the estrogen group presented higher figures than the control group, the IUD group was approximately equal, while in the progestagen groups the qualitative motility remained consistently decreased. In these groups there were naturally no motility results at 5 cm since the sperm had not reached this distance.

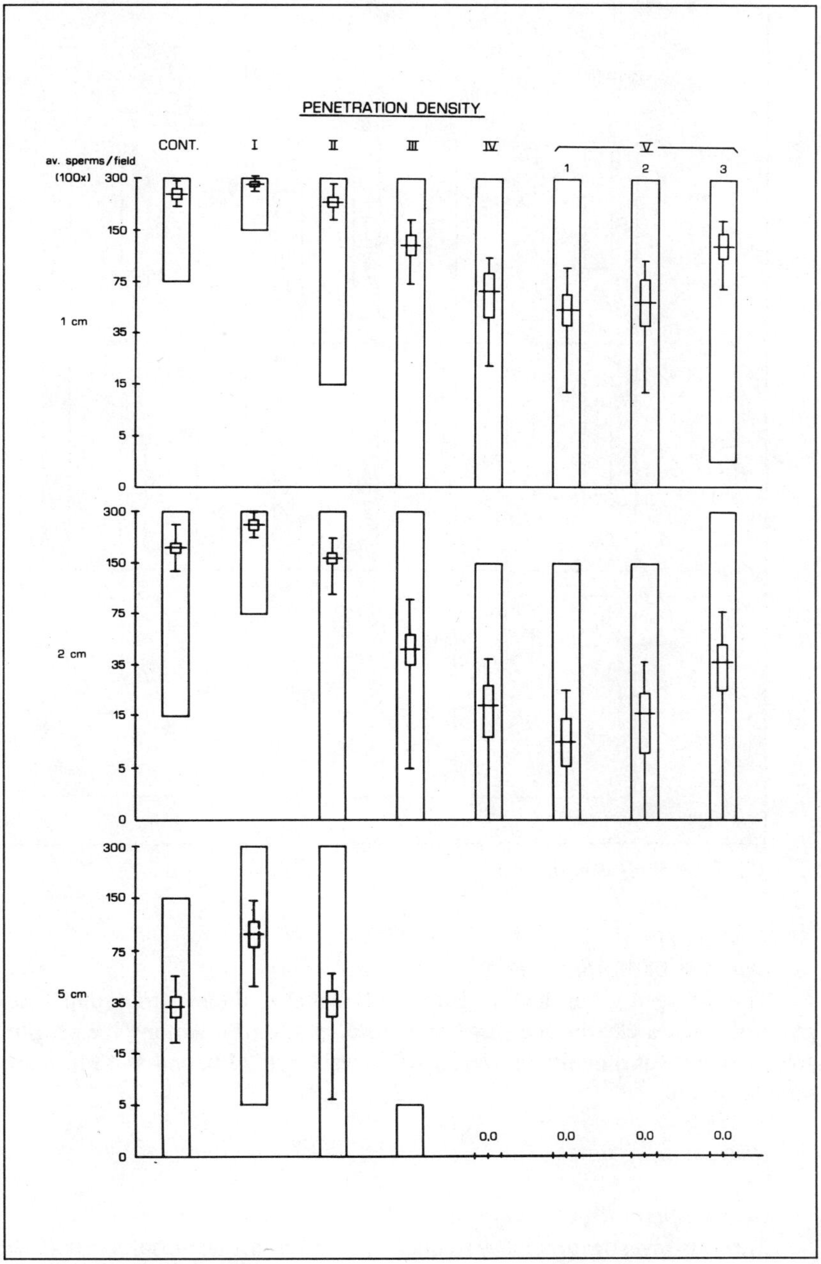

Fig. 6. Symbols as in figure 5.

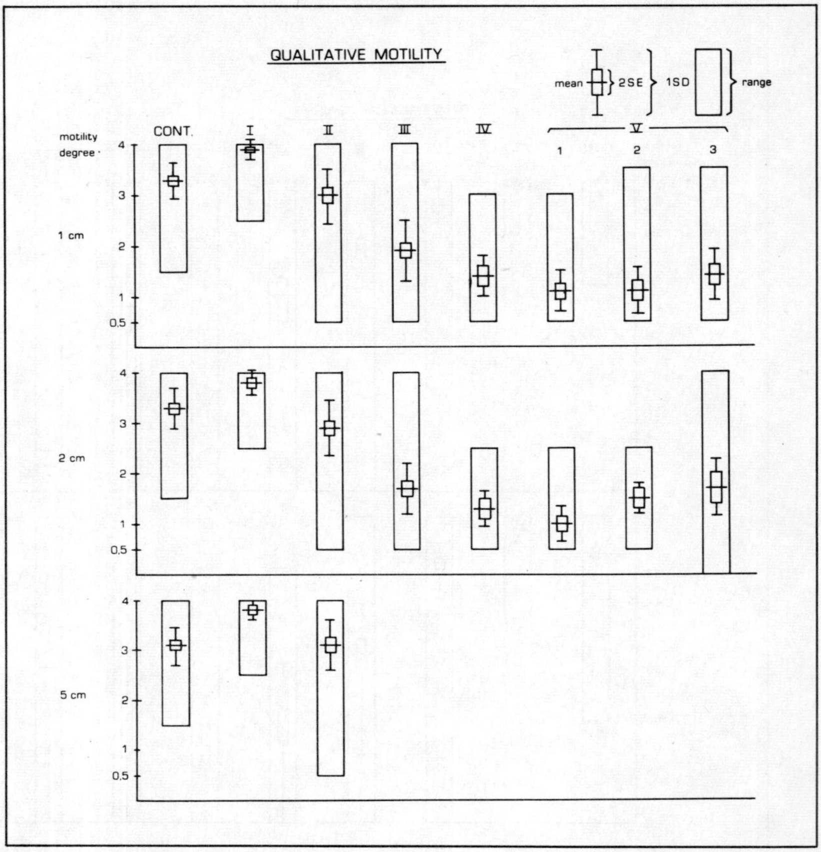

Fig. 7. Symbols as in figure 5.

Duration of Motility (fig. 8)

The estrogen group had a figure nearly equal to the control group. The IUD group showed a clearly decreased incidence ($p < 0.01$). Among the progestagen groups there was a marked lowering of the values at 24 h, and at 48 h, spermatic survival was nil.

Discussion

In vivo *Sperm Penetration*

Previous investigators have frequently expressed their results in a subjective or empirical manner. This has led to apparently different results and interpreta-

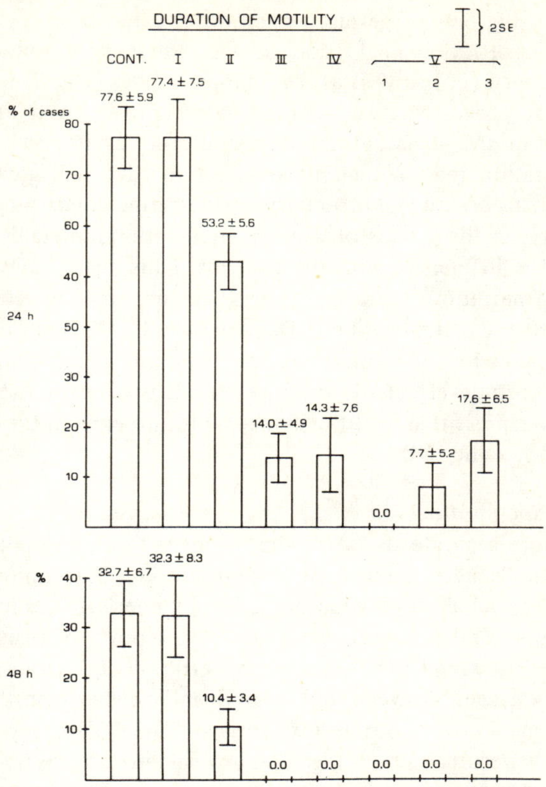

Fig. 8. Columns indicate percentages of each group with motile (living) sperm. Group symbols are in figure 5.

tions (3, 8, 14). In the present study, in order to make statistical evaluation possible, an attempt was made to obtain the number of sperm under conditions as uniform and rigorous as possible. Special attention has been devoted to the investigation of sperm ascent *beyond* the cervical canal, i.e. the uterine cavity; the actual information on this particular is scarce. To our knowledge, similar comparative studies are not available as yet. In the estrogen group our findings demonstrate that while intracervical penetration significantly improved with this treatment, the same did not occur in the intrauterine specimens. In the groups of women treated with contraceptives containing only progestagen, the results are most interesting. While intracervical penetration was not appreciably decreased in comparison to the control group, in the same subjects there was an almost *total absence* of sperm in the uterine specimens; the few exceptions showed very low counts. Using oral progestagen microdoses it has been reported that a good intracervical penetration might coexist with the complete absence of sperm in the uterine cavity (12) as well as the Fallopian tubes (10). Thus, it is possible to conclude that progestogen-containing contraceptives, even at minimal doses, constantly stop sperm ascent up to the uterine cavity, even though cervical sperm penetration may remain little affected. Although great variability was observed in intracervical and intrauterine penetration concerning the intrauterine contraceptives, in the groups of women fitted with nonmetallic IUDs there were no significative differences with the controls. Thus, these IUDs did not interfere with sperm penetration; there are reports with similar conclusions (1, 9). On the contrary, with the metal-bearing IUDs (Aries-Cu, Cu/Ag and Cu/Ni) we observed a definite suppression of intrauterine sperm ascent. This finding may be related to the spermiostatic effect of the three metals used, demonstrated *in vitro* (4, 13), and may represent an additional mechanism to explain the major contraceptive effect of metallic IUDs.

In vitro *Sperm Migration*

For a more accurate study of what happens to sperm in the cervical mucus, *in vitro* methods adequately complement postcoital techniques. The *Kremer* method used by us allows a good amount of numerical data to be obtained in a relatively simple and practical way. However, reports of results *in vitro* using contraceptives are scarce (7). Regarding penetration depth, our group of patients treated with estrogen showed figures significantly higher than the control group: this finding agrees with our results *in vivo.* The IUD group showed a wider variation in the results, as both high and low values were observed. This variation could be due to the mechanical disturbance of the characteristics of the mucus because of the presence of the IUD threads. The most important finding in this study was observed with the groups of patients on progestogen contraceptives. There is a *constant stop* of penetration that is produced at a distance of approximately 4 cm. This stop is observed even in the cases where penetration starts in an apparently normal way, similar to the control group. The results of the penetration density and qualitative motility corroborate the aforementioned. The duration of motility was the only parameter that did not improve with the estrogens and also the only one that became significantly worse with the IUDs. All of the progestagen groups showed markedly lower values, reaching absolute zero at 48 h. The explanation of the sperm migration 'stop' through mucus under the influence of a progestational effect still remains unknown. A reasonable suggestion is the theory of the molecular network (11) with a resulting mechanical blockage of sperm. However, further studies are needed in order to evaluate other possible processes involved, for instance, a specific biochemical action at the cervical level (5, 8).

References

1 *Bonney, W.A.; Glasser, S.R.; Clewe, T.H.; Noyes, R.W., and Cooper, C.L.:* Endometrial response to the intrauterine device. Amer. J. Obstet. Gynec. *96:* 101–110 (1966).

2 *Kesserü, E.:* Influence of various hormonal contraceptives on sperm migration *in vivo.* Fertil. Steril. *22:* 584–603 (1971).

3 *Kesserü, E.: In vivo* sperm penetration and *in vitro* sperm migration tests. Fertil. Steril. *28:* 584–591 (1973).

4 *Kesserü, E. and Camacho-Ortega, P.:* Influence of metals on *in vitro* sperm migration in the human cervical mucus. Contraception *6:* 231–240 (1972).

5 *Kesserü, E.; Camacho-Ortega, P.; Laudahn, G., and Schöpflin, G.: In vitro* action of progestagens on sperm migration in the human cervical mucus. IRCS Reprod. Obstet. Gynec. *2:* 1152 (1974).

6 *Kesserü, E. and Larrañaga, A.: In vitro* sperm migration in the human cervical mucus with different contraceptive methods. Contraception *3:* 195–208 (1971).

7 *Kremer, J.:* The *in vitro* spermatozoal penetration test in fertility investigations (Van Denderen, Groningen 1968).

8 *MacDonald, R.R. and Lumley, I.B.:* Sperm penetration of cervical mucus. Amer. J. Obstet. Gynec. *107:* 846–851 (1970).

9 *Malkani, P.K. and Sujan, S.:* Sperm migration in the female reproductive tract in the presence of intrauterine devices. Amer. J. Obstet. Gynec. *88:* 963–975 (1964).

10 *Martínez-Manautou, J.; Giner-Velázquez, J., and Rudel, H.W.:* Continuous progestogen contraception; a dose relationship study with chlormadinone acetate. Fertil. Steril. *18:* 52–72 (1967).

11 *Odeblad, E.:* Biophysics of the cervical mucus. Acta obstet. gynec. scand. *47:* suppl. 8, pp. 1–53 (1968).

12 *Roland, M.:* Norgestrel induced cervical barrier to sperm migration. J. Reprod. Fertil. Suppl. *5:* 173–177 (1968).

13 *Saito, S.; Bush, I.M., and Whitmore, W.F.:* Effects of certain metals and chelating agents on rat and dog epididymal spermatozoan motility. Fertil. Steril. *18:* 517–529 (1967).

14 *Zañartu, J.:* Effect of natural and synthetic sex steroids in cervical mucus, penetration and ascent of spermatozoa. Proc. 5th World Congr. Fertil. Steril., Stockholm 1966, pp. 704–709.

Author's address: Prof. *E. Kesserü*, Casilla 2236, *Lima* (Peru)

The Biology of Spermatozoa
INSERM Int. Symp., Nouzilly 1973, pp. 250–256 (Karger, Basel 1975)

Physiopathology of Sperm Transport in the Human Female

S. Jaszczak and E.S.E. Hafez

Departments of Gynecology and Obstetrics and Physiology, and C.S. Mott Center for Human Growth and Development, Wayne State University School of Medicine, Detroit, Mich.

Ejaculation

Semen is ejaculated on the cervix and posterior vaginal fornix. During midcycle, the cervical mucus protrudes from the exterior os, contacts with the whitish ejaculate which coagulates within 1 min, trapping most spermatozoa. Although the cervical mucus is immediately invaded by the sperm, there is at first no apparent mixture of the semen and cervical mucus (3). Fibrinolytic enzymes of the seminal plasma cause complete or partial liquefaction of the coagulum within 5–15 min (1, 2). The spermatozoa do not attain their full motility until liquefaction of the coagulum. The physiological significance of semen coagulation for sperm transport is not known.

During normal intercourse some 200–400 million spermatozoa are ejaculated. The first fraction of the ejaculate, however, contains the highest concentration of spermatozoa (4). However, cervical secretions which coat the upper part of the vagina and its fornices increase considerably the alkalinity of the vaginal milieu, and provide a favorable medium for spermatozoal activity. The number of sperm in a given ejaculate is an important factor for fertility, but the minimal number necessary to insure fertilization in man is not known.

Seminal plasma plays a major role in the transport and the physiology of spermatozoa. However, spermatozoa from the vas deferens and epididymis are successfully used in artificial insemination, and removal of various accessory organs of the male tract rarely decreases fertility as long as ejaculation still results in the release of a few million spermatozoa through the urethra (5).

Patterns of Sperm Transport in Female Tract

Transport in the Cervix

The endocervical mucosa is an intricate system of clefts or grooves grouped together. The endocervical canal contains approximately 100 secretory units, which secrete the bulk of mucus into the lumen of the canal (9). Several functions have been ascribed to the cervix and its secretion: (a) receptivity to sperm penetration at or near ovulation and inhibition of migration at other phases of the cycle, (b) sperm reservoir, (c) protection of spermatozoa from the hostile environment of vagina and from being phagocytized, (d) provision of spermatozoa energy requirements, (e) filtering of defective and immotile spermatozoa, and (f) possible capacitation of spermatozoa.

Ejaculated spermatozoa rapidly penetrate the watery cervical mucus during midcycle aided principally by both sperm motility and the micro- and macrorheological properties of mucus. The rate of sperm penetration in the mucus varies throughout the menstrual cycle from 0.1 to 3 mm/min and is maximal in the pre-ovulatory stage. Sperm penetrability of human cervical mucus begins approximately 8–5 days before ovulation and increases gradually to maximum at the time of ovulation. One or two days following ovulation sperm ascent decreases abruptly, but may persist to a lesser degree for an extended period of time (6). Individual variations between patients as well as between cycles of the same woman are common.

A highly significant relationship exists between the quantity, biochemical composition and physical properties of cervical mucus. Sperm penetration in cervical mucus is correlated with NaCl content and crystallization pattern of cervical mucus (7). *In vivo* and *in vitro* studies showed that when crystallization was below 40 %, spermatozoa penetrated a part of the cervical canal and did not reach the uterine cavity. They could not survive in the mucus for more than 24 h. When crystallization of cervical mucus was higher than 80 %, sperm migration from the cervical canal to the uterine cavity was associated with survival of sperm over 24 h.

Sperm passage through the endocervical canal results from (a) the intrinsic motility of spermatozoa and ability to overcome resistance of cervical mucus, (b) biochemical composition and physical properties of cervical mucus, and (c) contraction of the female reproductive tract as modified by sexual stimuli and pharmacologically active substances present in the semen.

Role of Sperm Motility

Early investigators believed that intercourse caused an increase in the muscular activity of the uterus, an inward-outward movement of the mucus column in the cervical canal with subsequent 'sucking in' of semen, to facilitate sperm migration. Thus, sperm transport was explained as a 'passive' process attributed to the negative pressure in the abdominal cavity of the female, independent of sperm motility. It seems unlikely that this negative pressure is effectively transmitted to the lower reproductive tract to enhance uterine contraction and sperm transport.

Patterns of Sperm Motility

The head of human spermatozoan measures 4.5 μm in length, 2.5 μm in width and 1.5 μm in thickness, which contains a large nucleus. Normal spermatozoa have an oval shaped head which is flattened in one dimension. Spermatozoa must be morphologically normal in order to migrate between the micelles of cervical mucus. Sperm migration through the cervical mucus is facilitated by the morphology of the sperm head which resembles a hydroplane. In order to maintain vigorous directional motility, the spermatozoon must overcome the resistance of the surrounding viscid medium (such as cervical mucus and endometrial fluid), and the resistance from the proximity of other spermatozoa, and the stiffness of its flagellum. Flagellation of the tail enables the sperm cell to penetrate the cervical mucus. The wave frequency of the propulsive force of spermatozoa seems to be provided by the entire flagellum and not by the distal segments of the tail. Little is known about the relationship between wave frequency and rate of rotation along the longitudinal exis of human spermatozoa.

Human spermatozoa show at least three types of motility: (a) progressive movement with nearly rectilinear resultant and small lateral head displacements; (b) progressive movement with some irregular changes of direction and generally large lateral head displacements, and (c) yawning, circular and helical motility (8). Under normal physiological conditions and in the fully activated state, spermatozoa show progressive movement, combined with a rotation along their longitudinal axis. The rotation is due to hydrodynamic effects arising from normal tail movements, consisting of a three-dimensional wave.

Highly active spermatozoa may beat synchronously in phase in suspensions where sperm concentration is dense. This observation while unconfirmed by other investigators, may explain the difference in the fertility indices between males who have 100 million normal sperm per ml as compared to those who have 20 million normal sperm per ml (3).

Filtering Phenomenon of Spermatozoa

Morphology of human spermatozoa varies in length, shape of head, and tail length. This is in contrast with the structural uniformity of spermatozoa in the semen of other mammals. The polymorphism of human spermatozoa, particularly of their heads, is so great that it is often difficult to define the morphological characteristics of a 'normal' spermatozoan. In a normal semen sample, 4–20 % of the spermatozoa have abnormal forms. Fertility seems to be impaired when the abnormal forms exceed 20 %.

There is a relationship between the morphology and penetrability of sperm through the cervical mucus (9). A 'filtering' phenomenon takes place when sperm penetrate into the micelles of cervical mucus since no abnormal forms penetrate the simulated endocervical canal although the semen contains several abnormal forms (9, 10).

Fractional postcoital tests were used to examine the relative distribution of sperm in the superior and inferior segments of the cervical canal. At 2 h postcoitum the lower part of the cervical canal contains a larger number of spermatozoa than the upper portion, but gradually sperm numbers increase in the upper part of the canal. At 6 h postcoitum the superior segment of the cervical canal contains a greater number of motile spermatozoa than the inferior segment (11). This 'filtering' effect of nonmotile sperm may be due to insufficient energy of abnormal sperm to travel through the cervical canal.

Physio-Pathological Factors Affecting Sperm Transport

pH Changes in Cervical Mucus

Vaginal content in women is usually acid (pH 3–4) and detrimental to ejaculated spermatozoa. Sperm migration and survival within the cervical canal are influenced by the pH of cervical mucus. The optimal pH for sperm ascent in the cervical mucus is between 7 and 8.5. There is a cyclic pattern of cervical mucus pH, with a distinct rise of alkalinity at the time of ovulation (7, 12). Spermatozoa are highly susceptible to changes in the pH of cervical mucus. Sperm mobility is reduced when the pH falls below 6.5 (13). The more alkaline pH of cervical mucus at the time of ovulation may favor sperm penetration and survival in the cervical canal and cervical crypts. Further studies are needed to determine the effect of pH on migration, survival and metabolism of spermatozoa. Although pH of cervical mucus is only one of several components of cervical mucus which favor sperm ascent at the time of ovulation, it cannot be underestimated in management of infertility. Infertile patients have higher chances to conceive if the pH of cervical mucus is alkaline. *In situ* measurement of pH of cervical mucus should be determined in evaluation of the cervix in infertile patients.

Cervical Pathology

There are large numbers of pathological conditions, which may be directly or indirectly related to the abnormal sperm migration through the cervical canal. These abnormalities may reduce fertility through several mechanisms such as inadequate secretion of cervical mucus ('dry cervix'), secretion of abnormal, hostile cervical mucus, malposition of the cervix and mechanical obstruction of the cervical canal.

Surgical procedures, lacerations and scarring following difficult deliveries, postabortum or postpartum excessive currettage, scarring following endocervical infections, or developmental abnormalities may decrease surface area of the cervical columnar epithelium.

Amputation, conization or electrocauterization of the cervix result in removal or destruction of large portions of cervical mucosa and decrease the amount of secreted mucus. The remaining of functional endocervical epithelium after these operations varies among patients as a result of the different depth of excision or electrocauterization. There are discrepancies in the literature and a paucity of studies regarding the effects of these procedures on subsequent fertility. Although most authors agree that a large proportion of patients fails to conceive after these operations, it must be stressed that several other factors may also be responsible for decreased fertility in these patients. These include: stenosis, synechiae, distortion and shortening of the cervix, natural decline in fertility with advancing age, disparenunia from associated vaginal surgery. Also, a large portion of such patients may not be attempting to become pregnant following these operations.

Cervical erosion (displacement of squamous by columnar epithelium) associated with cervicitis may play a role in infertility. The cervical mucus in such cases is contaminated with bacteria and leukocytes. When erosions are not infected, the cervical mucus is clean, clear, without significant numbers of bacteria and leukocytes, and with fairly good postcoital test. Cellularity and microbiological cultures should be estimated preferably at the time of expected ovulation in all patients with cervical erosions. Although the presence of bacteria in cervical mucus does not necessarily indicate that all of the microorganisms found are causing cervicitis, or that their presence requires antimicrobial therapy. Electrocauterization or conization in these cases may impair fertility as a result of destruction of mucus-secreting cervical crypts. If extensive Nabothian cysts are present near the external os, conization of the cervix may be beneficial (14).

Several microbial agents promote head-head agglutination of spermatozoa: *Escherichia coli* (15), myxoviruses (16), viruses of influenza, mumps and Newcastle disease (17) and *Pseudomonas maltophilia* (18). Mycoplasmas are also isolated from the cervical secretion of women with reproductive failures.

Spermicidal activity of *E. coli* and other bacteria *in vitro* has been shown. This effect can be prevented by the use of antibiotics resulting in 28 % of pregnancy and in 52 % improvement of postcoital tests. It has been reported that from 95 patients who harbored spermicidal organisms in the endocervical canal 22 % subsequently became pregnant after antimicrobial therapy.

Antigenic Properties of Semen

The antigenic properties of sperm and seminal fluid is well established. Several attempts have been made to show that seminal antigens delivered during coitus can induce an immune response in the female, and that the presence of antibodies to spermatozoa in the female can be correlated with infertility. The blood of 79 % of women with unexplained infertility contains sperm-agglutinating antibodies. However, other investigators report much lower values of 10–30 %.

Human cervical mucus contains IgG, IgA and IgM (19, 20). Using the immunohistologic technique IgG, IgA, and IgM have been found in the lumina of cervical crypts, basement membrane and interstitium of the cervix. The origin of immunoglobulins in cervical mucus is not clear, and their presence may result from transudation or from the local production within the cervix. In contrast to other external secretions the overall ratio of IgG to IgA in cervical mucus is 5.6 to 1 which is similar to blood serum.

The mobility and penetratability in cervical mucus of spermatozoa from men with autoantibodies or from men passively immunized with rabbit serum containing antibodies against seminal plasma and spermatozoa were extensively reduced (21). The degree of impairment of sperm mobility and penetration in cervical mucus was related to the titer of antibody. Low concentration of antibodies reduced sperm mobility. High concentration of antibodies inhibited both mobility and penetration of spermatozoa in cervical mucus (22).

Cervical and vaginal infections may stimulate local production of antispermatozoal antibodies or antibodies that have a close reaction with spermatozoa. Antibodies against spermatozoa in blood serum were detected in 63 % of 41 infertile patients with cervicovaginal schisostomiasis, as compared to 10 % of 20 (10 fertile and 10 infertile) patients with urinary or gastrointestinal forms and absence of antibodies in 14 fertile women with hepatic schisostomiasis. Antibodies detected in infertile women were antigen specific but not individually specific. A positive Franklin-Dukes test resulted when spermatozoa of husband as well as donors were used. Increased levels of IgA in blood serum of patients with cervicovaginal schisostomiasis was reported (23).

References

1 *Huggins, C. and Neal, W.:* Coagulation and liquefaction of semen: proteolytic enzymes and citrate in prostatic fluid. J. exp. Med. *76:* 527 (1942).
2 *Sobrero, A.J. and MacLeod, J.:* The immediate post-coital test. Fertil. Steril. *13:* 184 (1962).
3 *Davajan, V.; Nakamura, R.M., and Kharma, K.:* Spermatozoan transport in cervical mucus. Obstet. gynec. Surv. *25:* 1 (1970).
4 *MacLeod, J. and Hotchkiss, R.S.:* Distribution of spermatozoa and of certain chemical constituents in the human ejaculate. J. Urol., Baltimore *48:* 225 (1942).
5 *Noyes, R.W.:* Sperm transport; in *Behrman and Kistner* Progress in infertility, 1st ed. (Little, Brown, Boston 1968).
6 *Moghissi, K.S.:* Cyclic changes of cervical mucus in normal and progestin-treated women. Fertil. Steril. *17:* 663 (1966).
7 *MacDonald, R.R. and Lumley, L.B.:* Sperm penetration of cervical mucus. The role of saline. Amer. J. Obstet. Gynec. *107:* 847 (1970).
8 *Duijin, C. van, jr.; Voorst, C. van, and Freund, M.:* Movement characteristics of human spermatozoa analysed from kinemicrographs. Europ. J. Obstet. Gynec. *4:* 121 (1971).
9 *Botella-Llusia, J.:* Measurement of linear progression of the human spermatozoon as an index of male fertility. Int. J. Fertil. *1:* 113 (1956).
10 *Ruiz-Velasco, V.A.:* A quantitativespermatoc penetration test. Int. J. Fertil. *7:* 11 (1962).
11 *Davajan, V. and Kunitake, G.M.:* Fractional *in vivo* and *in vitro* examination of post-coital cervical mucus in the human. Fertil. Steril. *20:* 197 (1969).
12 *Moghissi, K.S.:* The function of the cervix in fertility. Fertil. Steril. *23:* 295 (1972).
13 *Parish, W.E. and Ward, A.:* Studies of cervical mucus from infertile women. J. Obstet. Gynaec. brit. Cwlth *75:* 1089 (1968).
14 *Roland, M.:* Norgestrel-induced cervical barrier to sperm migration. J. Reprod. Fertil. Suppl. *5:* 174 (1968).
15 *Taylor-Robinson, D. and Manchee, R.J.:* Spermadsorption and spermagglutination of mycoplasmas. Nature, Lond. *215:* 484 (1967).
16 *Peleg, B.A. and Ianconescu:* Spermagglutination and spermadsorption due to myxoviruses. Nature, Lond. *211:* 1211 (1966).
17 *Chu, H.P.:* The agglutination of spermatozoa by viruses of influenza, mumps, and Newcastle disease. Proc. 6th Int. Congr. Microbiol., vol. 3, p. 131 (1953).
18 *Bell, E.B.:* An immune-type agglutination of mouse spermatozoa by *Pseudomonas maltophilia.* J. Reprod. Fertil. *17:* 275 (1968).
19 *Schumacher, G.F.B.:* Biochemistry of cervical mucus. Fertil. Steril. *21:* 697 (1970).
20 *Behrman, S.J. and Lieberman, M.E.:* Biosynthesis of immunoglobulins by the human cervix; in *Blandau and Moghissi* Biology of the cervix (University of Chicago Press, Chicago 1973).
21 *Fjallbrant, B.:* Sperm antibodies and sterility in men. Acta obstet. ganec. scand. Suppl. *47:* 4 (1968).
22 *Shulman, S.:* Immunological barriers to fertility. Obstet. gynec. Surv. *27:* 553 (1972).
23 *Mahgoub, S.E.:* Antispermatozoal antibodies in infertile women with cervicovaginal schistosomiasis. A.J. Obstet. Gynecol. *112:* 781 (1972).

Authors' address: Dr. *S. Jaszczak* and Prof. *E.S.E. Hafez,* Department of Gynecology and Obstetrics, C.S. Mott Center for Human Growth and Development, Wayne State University School of Medicine, *Detroit, MI 48201* (USA)